ADVANCED
COMMON CORE
MATH
EXPLORATIONS

ADVANCED COMMON CORE MATH EXPLORATIONS

Fractions

JERRY BURKHART

PRUFROCK PRESS INC.

WACO, TEXAS

Prufrock Press Inc.
P.O. Box 8813
Waco, TX 76714-8813
Phone: (800) 998-2208
Fax: (800) 240-0333
http://www.prufrock.com

Table of Contents

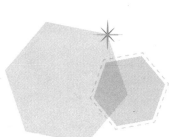

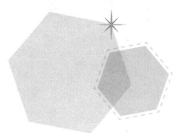

A Note to Students

Welcome, math explorers! You are about to embark on an adventure in learning. As you navigate the mathematical terrain in these activities, you will discover that "doing the math" means much more than calculating quickly and accurately. It means using your creativity and insight to question, investigate, describe, analyze, predict, and prove. It means venturing into unfamiliar territory, taking risks, and finding a way forward even when you're not sure which direction to go. And it means discovering things that will expand your mathematical imagination in entirely new directions.

Of course, the job of an explorer involves hard work. There may be times when it will take a real effort on your part to keep pushing forward. You may spend days or more pondering a single question or problem. Sometimes, you might even get completely lost. The process can be demanding—but it can also be very rewarding. There's nothing quite like the experience of making a breakthrough after a long stretch of hard work and seeing a world of new ideas and understandings open up before your eyes!

These explorations are challenging, so you might want to team up with a partner or two on your travels—to discuss plans and strategies and to share the rewards of your hard work. Even if you don't reach your final destination every time, I believe you will find that the journey was worth taking. So gear up for some adventure and hard work . . . and start exploring!

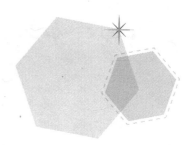

Introduction

This introduction contains general information about the structure of the books and activities in the Advanced Common Core Math Explorations series and how to use them. For additional information and support, please see the free e-booklet *Advanced Common Core Math Explorations: A Teacher's Guide* that accompanies this series at http://www.prufrock.com/Assets/ClientPages/AdvancedCCMath.aspx.

AUDIENCE

Advanced Common Core Math Explorations: Fractions is designed to support students, teachers, and other learners as they work to deepen their understanding of middle school math concepts. The activities have been written primarily with upper elementary and middle school students and teachers in mind. However, older students or those who have already studied more advanced content can also enjoy and benefit from them. The explorations can be used in classrooms, as professional development activities for mathematics teachers, in college math content and methods courses, and by anyone who would like to extend his or her understanding of middle school mathematics concepts by solving challenging problems.

These explorations are designed to stretch students beyond their initial level of comfort. They are built around the belief that most of us underestimate the mathematics we are capable of learning. Although the activities are challenging, they are also meant to be accessible. Although they are targeted to the special needs of gifted and talented students, I hope that teachers will make them available to any student who would like to pursue the challenge. Most students are capable of making progress and learning something meaningful, even if they work just on the first question or two of an activity.

PURPOSE

The investigations in this series were developed through years of work with talented middle school math students. They are designed to:
  » engage students in the excitement of mathematical discovery;
  » deepen students' understanding of a wide range of middle school math concepts;
  » encourage the use of multiple strategies for solving problems;

> » help students become flexible, creative, yet disciplined mathematical thinkers;
> » improve mathematical communication skills;
> » highlight connections between diverse mathematical concepts;
> » develop perseverance, patience, and stamina in solving mathematical problems;
> » provide levels of depth and challenge to meet a variety of needs and interests;
> » enable students to work both collaboratively and independently; and
> » offer opportunities for further exploration.

STRUCTURE OF THE BOOKS

The Advanced Common Core Math Explorations series contains ready-to-use explorations focused on one mathematical content area. The content and structure are built around the Common Core State Standards for Mathematics (National Governors Association Center for Best Practices & Council of Chief State School Officers [NGA & CCSSO], 2010), both the Content Standards and the Standards for Mathematical Practice. Because the emphasis is on challenge and depth, there is a stronger focus on concepts than on procedural skills. However, most activities provide plenty of opportunities to practice computational skills as well.

Each exploration is matched with one or more Common Core benchmarks or clusters, which come with grade-level designations (see p. 8). This grade-level information should serve as a rough guide. When selecting activities, use your own knowledge of your students' backgrounds and abilities. Information about the prior knowledge needed for each exploration is also included as a guide.

FEATURES OF THE EXPLORATIONS

Each activity includes three stages. Stage 1 (and sometimes part of Stage 2) may be challenging enough to meet the needs of many students. The second and third stages are usually appropriate for older students or for those who finish early, need more challenge, or are highly motivated and curious to learn more. They may also be useful for teachers or other adults who have more mathematical experience and want to extend their own knowledge further. I have separated the explorations into stages in order to provide a tool for setting goals, to help measure and celebrate students' progress, and to create additional options for those who need them.

Each exploration also contains features carefully designed to support teachers in the implementation process: an introduction, the student handout, a set of questions and notes to guide conversation, detailed solutions, and suggestions for a closing discussion.

IMPLEMENTING THE EXPLORATIONS

Implementing each exploration involves five steps on the part of the teacher: prepare, introduce, follow up, summarize, and assess.

Prepare

The best way to prepare to teach an activity is to try it yourself. Although this involves an initial time investment on your part, it pays great dividends later. Doing the activity, ideally with a partner or two, will help you become familiar with the mathematics, anticipate potential trouble spots for students, and plan ways to prepare students for success. After you have used the activity once or twice with students, very little preparation will be needed.

Introduce

The Introduction section at the beginning of each exploration provides support to help you get your students started, including materials and prior knowledge needed, learning goals, motivational background, and suggestions for launching the activity.

Read the Motivation and Purpose selection to students, and then follow the suggestions for leading a discussion to help them understand the problem. Often, one of the suggestions involves looking through the entire activity with them (or as much of it as they will be doing) to help them see the big picture before they begin. Let students know what time frame you have in mind for the exploration. An activity may take anywhere from a few days to 2 or 3 weeks depending on how challenging it is for students, how much of it they will complete, and how much class time will be devoted to it.

The explorations are designed to allow students to spend much of their time working without direct assistance. However, it's usually best if you stay with them for a few minutes just after introducing an activity to ensure that they get started successfully. This way, you can catch potential trouble spots early and prevent unnecessary discouragement.

This is also a good time to remind students about the importance of giving clear, thorough written explanations of their thinking. Specific motivation techniques and suggestions for developing mathematical communication skills are included in the *Advanced Common Core Math Explorations: A Teacher's Guide* e-booklet.

Follow Up

The level of challenge in these explorations makes it impractical for most students to complete them entirely on their own as seatwork or homework. Students' most meaningful (and enjoyable!) experiences are often the opportunities you give them to have mathematical conversations with you and with each other while the activity is in progress. If you are implementing an activity with a small group of students in a mainstream classroom, it may be sufficient to plan to meet with them a couple of times per week, for 15 or 20 minutes each time. If circumstances allow you more time than this, then the conversations and learning can be still better.

The Teacher's Guide for each exploration reprints each problem and contains two main elements: (a) Questions and Conversations and (b) Solutions. The Questions and Conversations feature is designed to help you facilitate conversations with and among students. For the most part, it lists questions that students may ask or that you may pose to them. Ideas for responding to the questions are included. It is not necessary to ask or answer all of the questions. Instead, let students' ideas and your experience and professional judgment determine the flow of the conversation. The Solution section offers ideas for follow-up discussions with students as they work. Although the answers in the Questions and Conversations sections are often intentionally incomplete or suggestive of ideas to consider, you will find detailed answers, often with samples of multiple approaches that students pursue, in the Solution section.

Summarize

After students have finished an exploration, plan a brief discussion (20 minutes is typically enough) to give them a chance to share and critique one another's ideas and strategies. This is also a good time to answer any remaining questions they have. The Wrap Up section at the end of each exploration offers ideas for this discussion, along with suggestions for further exploration.

Assess

One of the most valuable things you can do for your students is to comment on their work. You do not have to write a lot, but your comments should show that you have read and thought about what they have written. Whether you give praise or offer suggestions for growth, make your comments specific and sincere. Ideally, some of your comments will relate to the detail of the mathematical content. Some specific suggestions are included in the free e-booklet accompanying this series.

If you would like to give students a numerical score, consider using a rubric such as the one in *Extending the Challenge in Mathematics: Developing Mathematical Promise in K–8 Students* (Sheffield, 2003). Whatever system you use, the emphasis should be on process goals such as problem solving, reasoning, communication, and making connections—not just correct answers. You may also build in general criteria such as effort, perseverance, correct spelling and grammar, organization, legibility, etc. However, remember that the central goal is to develop students' mathematical capacity. Any scoring system should reflect this.

GETTING STARTED

Below are some tips for getting started. First, a few DON'Ts to help you avoid some common pitfalls:

» *Don't feel that you have to finish the activities.* Students will learn more from thinking deeply about one or two questions than from rushing to finish an activity. Each exploration is designed to contain problems that will challenge virtually any student. Most students will not be able to answer every question.

» *Don't feel that you have to explain everything to students.* Your most important job is to help them learn to develop and test their own ideas. They will learn more if they do most of the thinking.

» *Don't be afraid to allow students to struggle.* Talented students need to know that meaningful learning takes time and hard work. Many of them need to experience some frustration—and learn to manage it.

» *Don't feel that you have to know all of the answers.* In order to challenge our students mathematically, we have to do the same for ourselves. You'll never know all of the answers, but if you're like me, you'll learn more about the math every time you teach an exploration! Do what you can during the time you've allotted to prepare, and then allow yourself to learn from the mathematical conversations—right along with your students.

And now some important DOs:

» *Take your time.* Allow the students plenty of time to think about the problems. Take the time to explore the ideas in depth rather than rushing to get to the next question.

» *Play with the mathematics!* To many people's surprise, math is very much about creative play. Of course, there are learning goals, and it takes effort, but be sure to enjoy playing with the patterns, numbers, shapes, and ideas!

» *Listen closely to students' ideas and expect them to listen closely to each other.* Meaningful mathematical conversation may be the single most important key to students' learning. It is also your key to assessing their learning.

» *Help students feel comfortable taking risks.* When you place less emphasis on the answers and show more interest in the quality of students' engagement, ideas, creativity, and questions, they will feel freer to make mistakes and grow from them.

» *Believe that the students—and you—can do it!* Middle school students have great success with these activities, but it may take some time to adjust to the level of challenge.

» *Use the explorations flexibly.* You don't always have to use them exactly "as is." Feel free to insert, delete, or modify questions to meet your students' needs. Adjust due dates or completion goals as necessary based on your observations of students.

Many teachers find it helpful to make a solid but realistic commitment at the beginning of the school year to use the explorations. Put together a general plan for selecting students, forming groups, creating time for students to work (including time for you to meet with them), assessing the activities, and communicating with parents. Stick with your basic plan, making adjustments as needed as the school year progresses.

THE E-BOOKLET

The Advanced Common Core Math Explorations series comes with a free e-booklet (http://www.prufrock.com/Assets/ClientPages/AdvancedCCMath.aspx) that contains detailed suggestions and tools for bringing the activities to life in your classroom. It addresses topics such as motivation, questioning techniques, mathematical communication, assessment, parent communication, implementing the explorations in different settings, and identification.

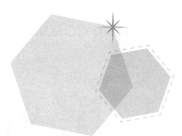

Connections to the Common Core State Standards

COMMON CORE STATE STANDARDS FOR MATHEMATICAL CONTENT

Table 1 outlines connections between the activities in *Advanced Common Core Math Explorations: Fractions* and the Common Core State Standards for Mathematics (NGA & CCSSO, 2010). The Standard column lists the CCSS Mathematical Content standards that apply to the activity. The Connections column shows other standards that are also addressed in the exploration. Extending the Core Learning describes how the activity extends student learning relative to the listed standard(s).

COMMON CORE STATE STANDARDS FOR MATHEMATICAL PRACTICE

The Common Core State Standards for Mathematical Practice are central to the purpose and structure of the activities in *Advanced Common Core Math Explorations: Fractions*. The list below outlines the ways in which the activities are built around these standards, providing a few specific examples for purposes of illustration.

1. **Make sense of problems and persevere in solving them.** All of the explorations in the *Advanced Common Core Math Explorations: Fractions* book engage students in understanding and solving problems. The process begins when you introduce the activity to your students and have a discussion in which everyone works together to clarify the meaning of the question and think about how to begin. Throughout each exploration, students devise problem-solving strategies, and make and test conjectures to guide their decisions and evaluate their progress as they work. They use visual models such as number lines and geometric drawings, and they create real-world stories to help them develop a deep understanding of the underlying concepts. To promote perseverance, the activities have a high level of cognitive demand, and there is support for the teacher and student in the form of motivation strategies, a tiered structure for the explorations, and suggestions for facilitating mathematical conversation.

TABLE 1

Alignment With Common Core State Standards for Mathematical Content

Exploration	Standard	Connections	Extending the Core Learning
1. Sharing and Grouping	5.NF.B.3 7.RP.A.1	6.NS.A.1	Relate simple and complex fractions to multiple meanings of division.
2. Fraction Puzzlers	5.NF.A 6.NS.C.6	4.NF.A 7.NS.A.1	Increase fluency and flexibility with concepts and procedures for adding, subtracting, and comparing fractions by solving challenging mathematical problems that have multiple solutions.
3. Working Together	5.NF.A.2 6.EE.A.2,3 7.EE.B.3	5.MD.A.1 6.RP.A	Develop a variety of strategies to solve challenging problems based on unit rates. Use patterns to generate algebraic expressions representing solutions and explain why they work.
4. Fractions Forever	5.NF.A.1 7.EE.B.3 8.NS.A.1	5.OA.A.2 5.OA.B.3, 7.NS.A.2	Interpret and create visual models of fractions to extend addition concepts to the case of infinite series. Discover and justify procedures for writing repeating decimals as fractions.
5. Visualizing Fraction Multiplication	5.NF.B		Create visual models and stories to develop, justify, and apply procedures for multiplying fractions and mixed numbers.
6. Undo It!	6.NS.A.1 5.NF.B.5 5.NF.B.7	5.NF.B.6 6.RP.A 7.RP.A.1	Understand the meaning of a reciprocal as an inverse for multiplication. Use the inverse relationship between multiplication and division to create, apply, and justify multiple strategies for dividing fractions.
7. Sum-Product Pairs	7.NS.A	7.EE.B	Increase fluency with adding, subtracting, multiplying, and dividing rational numbers while analyzing and extending complex patterns. Use algebraic expressions to represent and justify the patterns.
8. Unit Fraction Hunt	7.NS.A	7.EE.A.2 7.EE.B.3 8.EE.A.1 8.EE.A.2	Increase fluency with operations (including powers and roots) on rational numbers by solving challenging problems. Use properties of operations to create and write complex expressions in equivalent forms and to analyze the effects of making changes to the expressions.
9. Continued Fractions	7.NS.A.2 7.NS.A.3 8.NS.A 8.EE.A.2	7.EE.A.2 7.EE.B.3	Explore a system for representing numbers using reciprocals. Create procedures for translating between this system and fractions or decimals. Analyze and extend patterns within the system and use it to develop procedures for approximating irrational numbers.

2. **Reason abstractly and quantitatively.** The activities in this book provide students with frequent opportunities to understand and investigate connections between mathematical concepts and quantities, including fractions, division, reciprocals, rates, number properties, and algebra. The third exploration, "Working Together," provides a typical example. Here students create a visual representation to analyze a problem situation—paying close attention to the different units involved—and then develop algebraic expressions to formalize their solution processes. They return to the physical situation, testing their expressions against it and using them to gain new insights into the problem.

3. **Construct viable arguments and critique the reasoning of others.** These activities often prompt students to use what they have learned in earlier questions or explorations to justify a conclusion or explain why a new fact must be true. For example, in "Undo It!", students build on their knowledge of fraction multiplication models to devise multiple strategies for finding missing factors in fraction equations. They share methods and discuss their relative effectiveness, transparency, and efficiency. The "Questions and Conversations" and "Wrap Up" features in each exploration provide ongoing support for the teacher to lead these types of discussions.

4. **Model with mathematics.** In "Sharing and Grouping," "Visualizing Decimal Multiplication," and "Undo It!", students create visual models and real-world scenarios to match mathematical expressions and capture their meaning. They analyze one another's stories, striving to improve their correctness, appropriateness, and clarity.

5. **Use appropriate tools strategically.** Throughout these explorations, students use mental math, paper and pencil, graph paper, calculators, and visual models to solve problems. Occasionally, suggestions are provided as to the most effective tools. At other times, students are left to make their own decisions. In these cases, the advantages and disadvantages of one tool over another are discussed in the context of the problem situation and the learning goals for the activity.

6. **Attend to precision.** Students are consistently expected to give clear and complete explanations of strategies and procedures in these activities, both during discussion and in written work. They are expected to correctly use terms such as *numerator*, *denominator*, *reciprocal*, *unit rate*, and *rational number*. They carefully construct and label diagrams to capture the meanings of problem situations and numerical expressions. Teachers are also provided with support for leading discussions that develop students' communication skills. A section in the e-booklet accompanying the series is devoted to helping students understand why it is important to communicate clearly and precisely and how to do so effectively.

7. **Look for and make use of structure.** Pattern and structure are central components of the explorations in *Advanced Common Core Math Explorations: Fractions*. For example, in "Sum-Product Pairs," students

identify and extend patterns in computational processes and results, exploring the different forms they take as the numbers vary. In "Fraction Puzzlers," after solving one or two problems using any methods they can find, students have the opportunity to step back and look at previous results, finding ways to connect and adapt them to new situations.

8. **Look for and express regularity in repeated reasoning.** In *Advanced Common Core Math Explorations: Fractions*, students are constantly engaged in calculations and processes that display regularity. They use this predictability to find more efficient procedures, develop equations, and probe connections between concepts. For example, in "Fractions Forever!", students use visual models to explore infinite series. By noticing similarities in the processes and results, they discover new and more efficient approaches to carry out calculations. By rewriting numbers in decimal form and applying the patterns they have discovered, they develop new procedures for translating between decimal and fraction representations. In the final exploration of the book, students compare visual and symbolic forms of a new representation for numbers, continued fractions. They look for common features and use them to design algorithms to translate between continued fractions and traditional representations. By noticing patterns in their calculations, they also develop algebraic expressions to represent and simplify the procedures they have learned.

Exploration  1
Sharing and Grouping

INTRODUCTION

Materials

- » Graph paper (recommended)
- » Compasses and protractors (if students want to make circle models for fractions)

Prior Knowledge

- » Understand fractions in terms of equal parts of a whole.
- » Understand equivalent fractions, and use them to add and subtract fractions.
- » Know the meanings of *dividend*, *divisor*, and *quotient*.

Note. This exploration works best if you use it before students have learned procedures for multiplying and dividing fractions.

Learning Goals

- » Represent equivalent fractions and division with visual models.
- » Create and analyze real-world situations that connect division and fractions.
- » Apply "How many groups?", "How many in each group?", and comparison models of division to cases in which the quotient, dividend, or divisor are fractions.
- » Explore meanings and strategies for simplifying complex fractions.
- » Communicate complex mathematical ideas clearly.
- » Persist in solving challenging problems.

Launching the Exploration

Motivation and purpose. To students: Division and fractions have a variety of meanings. In this exploration, you will create and solve real-world problems in order to understand some of these meanings. In Stages 2 and 3, you will extend your knowledge to complex fractions—fractions whose numerators or denominators are fractions themselves!

Understanding the problem. Discuss different meanings of division with students. Ask them to provide examples of each using whole numbers:

11

» In "How many (or much) in each group?" division, you know the number of groups and want to know the size of each group.

» In "How many groups?" division, you know how many or how much in each group and want to know how many groups there are.

» Closely related to both of these is the *comparison* model of division in which you ask "How many times as much?" or "What fraction as much?"

Teacher's Note. These descriptions of the meanings of division come from the Common Core State Standards for Mathematics (NGA & CCSSO, 2010, pg. 89). "How many in each group?" division is also called *partitive* division, while "How many groups?" division is sometimes referred to as *quotative* (or *measurement*) division.

Tell students that this activity is about investigating connections between division and fractions. In particular, it is about exploring these meanings of division in situations in which the dividend, divisor, or quotient is a fraction.

Look through the entire activity with students (or as much of it as they will be doing) to help them see how everything fits together. Stage 1 is about solving problems and creating stories to explore division in situations in which the quotient is a fraction. In Stages 2 and 3, students learn how to understand and work with complex fractions, in which the numerator or denominator itself is a fraction!

Make sure students understand Problem #1. Emphasize that the goal is to find the fraction of a sandwich for each person.

NAME: _____ DATE: _____

STUDENT HANDOUT

Stage 1

1. You and six of your friends are visiting a submarine sandwich shop. Between the 7 of you, there is enough money to buy 3 subs. You are all hungry, but being good friends you want to share them equally. Which meaning of division does this story illustrate? What fraction of a sub does each person get? Draw a diagram to illustrate your solution and your strategy. Explain.

2. Repeat the previous question for at least one more case in which the number of people is greater than the number of subs. Write a number model for each situation. Use your number models to make a conjecture. Is your conjecture still true when the number of subs is greater than the number of people? Explain.

3. Now it's your turn to create a story! Use the "How many groups?" meaning of division to write a story problem to fit the expression $24 \div 6$. Does your story still make sense for $24 \div 15$? How about $24 \div 42$? If not, change it so that it works well for these expressions, too. Explain your thinking.

4. Draw diagrams to illustrate $24 \div 15$ and $24 \div 42$. Explain how the diagrams show the answers and fit the meaning of your story problem.

Stage 2

The numbers in Problems #5 and #6 are called *complex fractions* because they contain fractions in the numerator or denominator (or both). Fractions that have whole numbers in the numerator and denominator are known as *simple fractions*.

5. Draw a diagram to represent the fraction $\dfrac{1\frac{1}{2}}{5}$. Create a story or situation that this fraction could describe. Then explain how to use the diagram to help you write it as an equivalent simple fraction.

6. Repeat the process in Problem #5 for the fraction $\dfrac{2}{3\frac{1}{3}}$.

7. You are making cookies from a recipe that calls for $\dfrac{1}{3}$ of a cup of butter. One stick of butter contains $\dfrac{1}{2}$ of a cup. Write a complex fraction to represent this situation. Draw a diagram, and show how to use it to write the complex fraction as an equivalent simple fraction. What fraction of the stick do you need?

8. Invent a method to write a complex fraction as an equivalent simple fraction without using diagrams. Apply your method to the complex fractions in Problems #5, #6, and #7.

Stage 3

9. Draw a diagram to represent the complex fraction $\dfrac{1\frac{3}{4}}{2\frac{1}{2}}$. Create a story or situation that this fraction could represent. Explain how to use your diagram to write this as a simple fraction. Then show how you could do this without the diagram!

TEACHER'S GUIDE

STAGE 1

Problem #1

1. You and six of your friends are visiting a submarine sandwich shop. Between the 7 of you, there is enough money to buy 3 subs. You are all hungry, but being good friends you want to share them equally. Which meaning of division does this story illustrate? What fraction of a sub does each person get? Draw a diagram to illustrate your solution and your strategy. Explain.

Questions and Conversations for #1

This section contains ideas for conversations, mainly in the form of questions that students may ask or that you may pose to them. Be sure to allow students to do most of the thinking and talking!

» *What would the answer be if there were 6 people instead of 7? Will the answer to the original question be greater or less than this? Why?* For 6 people, the answer would be $\frac{1}{2}$. The answer to the original question will be less than this because you are sharing between more people.

> **Teacher's Note.** Allow students plenty of time to work independently before intervening to discuss the first question. They may create strategies that surprise you!

» *Does it make more sense to express your answer in decimal or fraction form? Why?* You may be tempted to use long division to find an answer in decimal form. However, the fraction form is easier to visualize. Also, notice that the question asks for the fraction of a sub.

» *Into how many parts should you split each sub?* Seven parts works best for most people, but allow students to experiment with other possibilities. Ultimately, the goal is for them to understand the general strategy of dividing each sandwich into a number of parts equal to the number of people who are sharing them.

» *Why isn't the answer $\frac{3}{21}$? (This question is for students who have drawn a good picture but are interpreting it incorrectly.)* Reread the question and think carefully: What is the "whole"?

> **Teacher's Note.** Students who make this error may not understand that the denominator represents the number of parts in *one whole*. They may also mistakenly believe that all parts must come from the same whole.

Solution for #1

This story uses the "How much in each group?" meaning of division. The answer is $\frac{3}{7}$ of a sub. There are many strategies. The most effective ones usually involve splitting each sub into 7 parts.

Sample response 1:

1	1	1	2	2	2	3

3	3	4	4	4	5	5

5	6	6	6	7	7	7

There are 21 parts all together in the 3 subs. Because you are sharing between three people, each person receives $21 \div 7 = 3$ of these parts. Because each part is $\frac{1}{7}$ of a sub, each person gets $\frac{1}{7} + \frac{1}{7} + \frac{1}{7} = \frac{3}{7}$ of a sub. (Students often imagine portioning them out as shown on the left.)

Sample response 2:

1	2	3	4	5	6	7

1	2	3	4	5	6	7

1	2	3	4	5	6	7

Each person could get $\frac{1}{7}$ of each sub. This method is easier for students to generalize to other numbers, because they can see the solution at a glance instead of counting the total number of parts.

Sample response 3: Here's an example of a creative one!

1	2	3

4	5	6

7	1 2 3 4 5 6 7

Split the subs into thirds. Give one of the thirds to each person. This leaves $\frac{2}{3}$ of a sub remaining, which must also be shared equally among the 7 people.

At this point, it starts to get tricky! If students struggle for a while, consider steering them toward a simpler approach. If they want to follow through with their idea, suggest that they break the 7 small pieces into smaller pieces (two parts each works well). See if they can make it work from there!

Sample response 4: Some students may find the answer before they draw a diagram. They might recognize that it is a division situation, and they may already know that $3 \div 7 = \frac{3}{7}$. This is great! However, these students should be able to prove their answer with a drawing.

Problem #2

2. Repeat the previous question for at least one more case in which the number of people is greater than the number of subs. Write a number model for each situation. Use your number models to make a conjecture. Is your conjecture still true when the number of subs is greater than the number of people? Explain.

Questions and Conversations for #2

» *What is a number model?* In this case, you can think of it as an equation.

» *Are there patterns in your number models?* There should be. Compare the dividend and divisor to the answer in fraction form.

» *Can you express these patterns using words or algebraic symbols?* Yes, you can do both.

> **Teacher's Note.** Students may have trouble seeing the pattern if they are simplifying their fractions. Suggest that they choose some values of a and b that have a greatest common factor of 1.

Solution for #2

Below is an answer for 5 subs and 9 people. The result is $\frac{5}{9}$ of a sub for each person. This example illustrates only the second strategy, but students may have other approaches.

1	2	3	4	5	6	7	8	9
1	2	3	4	5	6	7	8	9
1	2	3	4	5	6	7	8	9
1	2	3	4	5	6	7	8	9
1	2	3	4	5	6	7	8	9

Split each sub into 9 equal parts. Each person receives $\frac{1}{9}$ of each sub for a total of $\frac{1}{9}+\frac{1}{9}+\frac{1}{9}+\frac{1}{9}+\frac{1}{9}=\frac{5}{9}$ of a sub.

Number models for the two previous examples are $3 \div 7 = \frac{3}{7}$ and $5 \div 9 = \frac{5}{9}$. Students may use these to conjecture that when you express a quotient as a fraction, the dividend becomes the numerator and the divisor becomes the denominator. The equation $a \div b = \frac{a}{b}$ expresses this relationship algebraically.

> **Teacher's Note.** Regardless of the strategy that students initially use, make sure they understand that when you share between n people, it works best to divide each sub into n equal parts. This strategy is both transparent and effective for "How many in each group?" division. Students can readily see that if you divide 1 into b equal groups, each group contains $\frac{1}{b}$. Therefore, when you divide a into b equal groups, each part contains a groups of $\frac{1}{b}$, which is $\frac{a}{b}$.

Problem #3

3. Now it's your turn to create a story! Use the "How many groups?" meaning of division to write a story problem to fit the expression $24 \div 6$. Does your story still make sense for $24 \div 15$? How about $24 \div 42$? If not, change it so that it works well for these expressions, too. Explain your thinking.

Questions and Conversations for #3

  » *Why do stories that work for $24 \div 6$ not always work for $24 \div 15$ and $24 \div 42$?* $24 \div 6$ represents a whole number, while the other two do not. Also, $24 \div 42$ is less than 1. Mixed numbers and numbers less than 1 may not make sense in all cases.

Solution for #3

Sample story 1: Alicia earns $6 per hour. How many hours does it take her to earn $24? This story fits the "How many groups?" meaning of division because it involves finding the number of groups of 6 in 24. The story also makes sense for $24 \div 15$ and $24 \div 42$:

  » If Alicia earns $15 per hour, she works $24 \div 15 = 1\frac{3}{5}$ hours to earn $24.

  » If Alicia earns $42 per hour, she works $24 \div 42 = \frac{4}{7}$ of an hour to earn $24.

Sample story 2: A gymnastics school owns vans that hold 6 people each. How many vans are needed to transport 24 people to a meet?

If you want the fractional answers to make sense for $24 \div 15$ and $24 \div 42$, you might change the question from "How many vans are needed . . . ?" to "How many buses will be filled . . . ?" This allows you to think in terms of filling part of a bus. (And buses are more likely than vans to hold a larger number of students.)

  » If a bus holds 15 people, it will fill $24 \div 15 = 1\frac{3}{5}$ buses when you transport 24 people.
  » If a bus holds 42 people, it will fill $24 \div 42 = \frac{4}{7}$ of a bus when you transport 24 people.

Note. Some students may have to change their question entirely in order to make it work in all three cases.

Problem #4

4. Draw diagrams to illustrate $24 \div 15$ and $24 \div 42$. Explain how the diagrams show the answers and fit the meaning of your story problem.

Questions and Conversations for #4

» *Should you write your quotients in fraction or decimal form?* Write them in fraction (or mixed number) form so that you can explore connections between division and fractions. You may also write them in decimal form if you like, but some of the decimals might get messy!

» *Should you show fractions in simplest form?* Yes. Try to see how the simplest form shows up in your drawings.

Solution for #4

Sample diagram showing that 24 is $1\frac{3}{5}$ groups of 15:

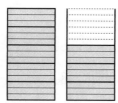 Each horizontal bar stands for one student. The 24 students fill one entire bus and 9 of the 15 places in the second bus, or $1\frac{9}{15}$ buses. The dark dividing lines show groups of 3 students, each of which fills $\frac{1}{5}$ of a bus. Looking at these, we see that $1\frac{9}{15}$ is equivalent to $1\frac{3}{5}$ buses.

Sample diagram showing that 24 is $\frac{4}{7}$ of a group of 42:

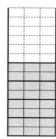 Each small rectangle stands for a student. The diagram shows that 24 students fill $\frac{24}{42}$ of a bus that holds 42 students. The dark dividing lines show groups of 6 students, each of which fills $\frac{1}{7}$ of the bus. Looking at these, we can see that $\frac{24}{42}$ is equivalent to $\frac{4}{7}$ of a bus.

Notice how "How many groups?" division looks like the "number of equal parts in a whole" interpretation of fractions when the quotient is less than 1.

STAGE 2

The numbers in Problems 5 and 6 are called *complex fractions* because they contain fractions in the numerator or denominator (or both). Fractions that have whole numbers in the numerator and denominator are known as *simple fractions*.

Problem #5

5. Draw a diagram to represent the fraction $\dfrac{1\frac{1}{2}}{5}$. Create a story or situation that this fraction could describe. Then explain how to use the diagram to help you write it as an equivalent simple fraction.

Questions and Conversations for #5

» *Is a simple fraction the same thing as a fraction in simplest form?* Not quite. For example, $\frac{6}{9}$ is a simple fraction because the numerator and denominator are whole numbers. However, it is not in simplest form—it can be simplified to $\frac{2}{3}$.

» *Does it matter what type of a diagram you use to show the complex fractions?* There are several types of diagrams you can use to represent fractions. Most people use rectangles, circles, or number lines. Consider using a variety of diagrams in this exploration!

» $\frac{1\frac{1}{2}}{5}$ *is between what two simple fractions?* The most natural choices are $\frac{1}{5}$ and $\frac{2}{5}$. Into how many parts should you split the whole in order to show $1\frac{1}{2}$ fifths?

Solution for #5

Sample diagram and story:

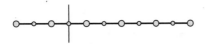

You are riding your bike to see a friend who lives 5 miles away. You have traveled $1\frac{1}{2}$ miles so far. What fraction of the trip have you completed?

The large dots mark off the miles. You can write $\frac{1\frac{1}{2}}{5}$ as a simple fraction by splitting each of the 5 miles into 2 equal parts, creating 10 equal parts in the whole. You have traveled 3 of the 10 equal parts, which represents $\frac{3}{10}$.

Problem #6

6. Repeat the process in Problem #5 for the fraction $\frac{2}{3\frac{1}{3}}$.

Questions and Conversations for #6

» *How many parts are in the whole? How can you show this?* Because the denominator is $3\frac{1}{3}$, there are $3\frac{1}{3}$ parts in the whole. Into how many equal-size parts can you split the whole in order to show this clearly?

Solution for #6

Sample diagram and story:

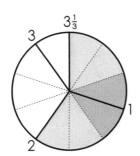

You are watching an epic movie that lasts $3\frac{1}{3}$ hours. You have seen 2 hours of it so far. What fraction of the movie have you watched?

You can write $\dfrac{2}{3\frac{1}{3}}$ as a simple fraction by splitting each of the 3 units into 3 equal parts. Taking account of the final $\frac{1}{3}$ of a part, this creates 10 smaller equal parts. Now there are 6 parts out of 10, which represents $\dfrac{6}{10}$ or $\dfrac{3}{5}$.

Problem #7

7. You are making cookies from a recipe that calls for $\frac{1}{3}$ of a cup of butter. One stick of butter contains $\frac{1}{2}$ of a cup. Write a complex fraction to represent this situation. Draw a diagram, and show how to use it to write the complex fraction as an equivalent simple fraction. What fraction of the stick do you need?

Questions and Conversations for #7

» *Can you add anything to your drawing of the stick of butter to make the situation easier to visualize or understand?* Some people like to show two sticks of butter. Can you see why?

Solution for #7

The diagram shows $\frac{1}{3}$ out of $\frac{1}{2}$ cups of butter, or $\dfrac{\frac{1}{3}}{\frac{1}{2}}$.

Measurements in cups:

The stick of butter is outlined in bold. Some students may add an "imaginary" stick (shown to the right with dotted lines) so that they can compare to a whole cup. The diagram shows that $\frac{1}{3}$ of a cup of butter is $\frac{2}{3}$ of a stick.

Problem #8

8. Invent a method to write a complex fraction as an equivalent simple fraction without using diagrams. Apply your method to the complex fractions in Problems #5, #6, and #7.

Questions and Conversations for #8

» *In Problems #5–#7, how do the numerators and denominators of the simple fractions relate to those of the original fractions?* Compare their sizes and look for patterns.

» *Can you apply your existing knowledge of equivalent fractions?* Think of the procedures you have learned for creating equivalent fractions. Test them to see if they still make sense for complex fractions.

Solution for #8

In Problem #5, you split each mile into two equal parts, which doubled the numerator and denominator. In Problem #6, you split each hour into three equal parts, which tripled the numerator and denominator. This suggests that you can use the usual process of multiplying the numerator and denominator by the same number:

$$\frac{1\frac{1}{2}}{5} = \frac{1\frac{1}{2} \cdot 2}{5 \cdot 2} = \frac{3}{10} \qquad \frac{2}{3\frac{1}{3}} = \frac{2 \cdot 3}{3\frac{1}{3} \cdot 3} = \frac{6}{10} = \frac{3}{5}$$

This works for Problem #7 as well. Multiply the numerator and denominator by a number (their least common multiple works well) that will make them both whole numbers:

$$\frac{\frac{1}{3}}{\frac{1}{2}} = \frac{\frac{1}{3} \cdot 6}{\frac{1}{2} \cdot 6} = \frac{2}{3}$$

Teacher's Note. Students need not have been taught procedures for multiplying fractions or mixed numbers to do this. They can use repeated addition or strategies of their own.

STAGE 3

Problem #9

9. Draw a diagram to represent the complex fraction $\dfrac{1\frac{3}{4}}{2\frac{1}{2}}$. Create a story or situation that this fraction could represent. Explain how to use your dia-

gram to write this as a simple fraction. Then show how you could do this without the diagram!

Questions and Conversations for #9

»   *What type of diagram works best for a complex fraction that has mixed numbers in both the numerator and denominator?* You may be in the habit of drawing circles to represent fractions of all types. Experiment and decide what works best for you, but you might find that a number line makes the fraction easier to visualize, especially when the complex fraction contains mixed numbers.

Solution for #9

Sample diagram and story:

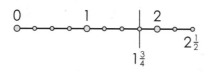

A bag can hold $2\frac{1}{2}$ pounds of flour. If you pour $1\frac{3}{4}$ pounds of flour into the empty bag, what fraction of its full weight is it holding?

It is holding $\dfrac{1\frac{3}{4}}{2\frac{1}{2}}$ of its full weight. The simplified value of this complex fraction is $\dfrac{7}{10}$, because the diagram shows that $1\frac{3}{4}$ is $\dfrac{7}{10}$ of the way to $2\frac{1}{2}$.

The complex fraction can also be simplified by multiplying the numerator and denominator by 4. (Notice that this is the same number of parts into which each unit was split in the diagram.)

$$\frac{1\frac{3}{4}}{2\frac{1}{2}} = \frac{1\frac{3}{4} \cdot 4}{2\frac{1}{2} \cdot 4} = \frac{7}{10}$$

WRAP UP

Share Strategies

Have students share and critique one another's diagrams, stories, and calculation strategies.

Summarize

Answer any remaining questions that students have. Summarize some key ideas: Two important meanings for division are based on asking "How many groups?" and "How many in each group?" These two meanings of division and the language we use to describe them can appear to change in subtle ways when fractions and mixed numbers are present, but the underlying ideas remain the same. In either case, a key relationship between fractions and division can be expressed as $a \div b = \dfrac{a}{b}$. (Some students may have already learned to apply this relationship when using division to write fractions in decimal form. This exploration shows that the concept goes much deeper than this.)

Further Exploration

Ask students to think of new questions to ask or ways to extend this exploration. Here are some possibilities:

» Can you create "How many in each group?" stories and diagrams for cases in which the divisor is a fraction or mixed number?

» What happens when you have a complex fraction with the number 1 in the numerator? Describe any patterns you see and use diagrams and stories to explain why they occur.

» Design diagrams and stories for complex fractions such as $\dfrac{2}{\frac{3}{4}}$ or $\dfrac{\frac{2}{3}}{\frac{3}{7}}$, in which the numerator is greater than the denominator.

» Create and explore some extended complex fractions, such as $\dfrac{\frac{1}{2}}{\frac{4}{\frac{2}{5}}}$.

Can you design procedures for writing these as simple fractions? Is it feasible to create diagrams and stories for them?

Exploration 2
Fraction Puzzlers

INTRODUCTION

Prior Knowledge

- » Understand the meaning of equivalent fractions.
- » Know procedures for adding and subtracting fractions.
- » Locate fractions on a number line.

Learning Goals

- » Think flexibly about fractions.
- » Find multiple strategies to solve problems that involve equivalence, comparisons, sums, and differences of fractions.
- » Increase fluency with procedures for comparing, adding, and subtracting fractions.
- » Organize data to recognize the point at which all solutions have been found (Stage 3).
- » Communicate complex mathematical ideas clearly.
- » Persist in solving challenging problems.

Launching the Exploration

Motivation and purpose. To students: In this fraction exploration, you are given the answer and asked to find the questions! Because there may be many solutions—and even more ways to find them—you will need to apply creativity and ingenuity. You will finish with a deeper understanding of how fractions work.

Understanding the problem. Students will best achieve the learning goals of this activity by solving the problems using fractions (not decimals) and by using calculators sparingly, if at all. Of course, at some other time, students may benefit from experimenting with different approaches.

Read through the first page containing the instructions for Stage 1 and answer students' questions. Then skim through the entire activity (or the part that students will be completing). The problem statements are short, but they may take a quite a while to solve. Tell students not to get discouraged if they are not able to find solutions quickly. The process may speed up once they have worked for a while and have begun to develop strategies.

As students begin working on the first question, check that they understand the conditions of the problem, including the fact that A, B, and C must all lie in the range from 0 to 1. This reduces the number of solutions and makes it more practical for them to use the number line as part of their thinking process. The number line helps them focus their attention on the size of each fraction—something that is easy to forget when they are caught up in the details of the calculations.

Teacher's Note. This exploration might be quite challenging for students (and teachers!)—even more than some later activities in the book. However, developing flexible reasoning and computation skills with fraction addition, subtraction, and comparison is important for students' future learning. Be sure to allow them plenty of time to think about these questions. They may struggle at the beginning, but, with sufficient time and support, they will be successful!

NAME: _____ DATE: _____

STUDENT HANDOUT

Stage 1

For each of the Problems #1–#5, you have an equation template and a number line:

You are given a number and asked to create it by filling in the numerator and denominator of each fraction with 1, 2, 3, 4, 5, or 6. You must use each number exactly once, and you may choose either addition or subtraction between each pair of fractions.

Then, on the number line, locate and label "A" the starting fraction, "B" the sum or difference of the first two fractions, and "C" the final answer. None of these three numbers may ever be less than 0 or greater than 1.

Example: Create the number $\frac{4}{15}$.

$$\frac{1}{6} + \frac{3}{5} - \frac{2}{4} = \frac{4}{15}$$

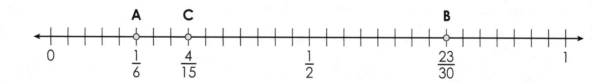

"A" is located at $\frac{1}{6}$, the fraction you started with. "B" is at $\frac{23}{30}$, the sum of the first two fractions, $\frac{1}{6} + \frac{3}{5}$. "C" shows the final result, $\frac{4}{15}$.

Stage 1

Using the instructions on the previous page, show your answers to Problems #1–#5 on the recording sheet.

1. Create the number 0.
2. Create the number 1.
3. Create a number as close as possible to but not equal to $\frac{1}{2}$.
4. Create a number as close as possible to but not equal to 0.
5. Create a number as close as possible to but not equal to 1.

Stage 2

6. Find at least 10 solutions to the equation below. a, b, and c must be whole numbers. Describe your strategies. Show the calculations for at least five of your solutions to prove that they are correct.

$$\frac{1}{a} + \frac{1}{b} + \frac{1}{c} = \frac{1}{5}$$

Stage 3

7. Find all solutions to the equation in Problem #6. Explain how you can be sure you have found them all. (a, b, and c must still be whole numbers.)

FRACTION PUZZLERS: RECORDING SHEET

1.

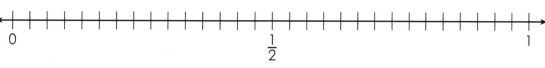

2.

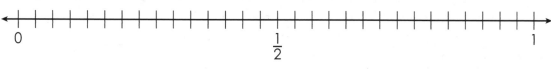

3.

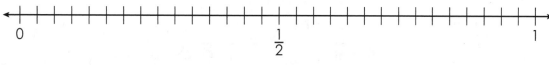

4.

5.

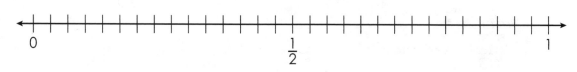

TEACHER'S GUIDE

STAGE 1

For each of the Problems #1–#5, you have an equation template and a number line:

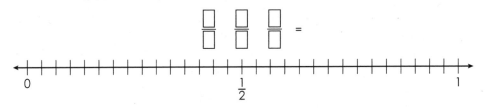

You are given a number and asked to create it by filling in the numerator and denominator of each fraction with 1, 2, 3, 4, 5, or 6. You must use each number exactly once and you may choose either addition or subtraction between each pair of fractions.

Then, on the number line, locate and label "A" the starting fraction, "B" the sum or difference of the first two fractions, and "C" the final answer. None of these three numbers may ever be less than 0 or greater than 1.

Example: Create the number $\dfrac{4}{15}$.

$$\boxed{\dfrac{1}{6}} + \boxed{\dfrac{3}{5}} - \boxed{\dfrac{2}{4}} = \dfrac{4}{15}$$

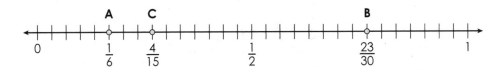

"A" is located at $\dfrac{1}{6}$, the fraction you started with. "B" is at $\dfrac{23}{30}$, the sum of the first two fractions, $\dfrac{1}{6} + \dfrac{3}{5}$. "C" shows the final result, $\dfrac{4}{15}$.

Using the instructions, show your answers to Problems #1–#5 on the recording sheet.

Problem #1

1. Create the number 0.

Questions and Conversations for #1

This section contains ideas for conversations, mainly in the form of questions that students may ask or that you may pose to them. Be sure to allow students to do most of the thinking and talking!

» *Can you think of ways to transform unsuccessful attempts into correct solutions—or old solutions into new ones?* Both of these are possible. Never underestimate what you can learn from things that did not work the first time! Always look at old solutions to see if they lead to new ones. For example, you might look at a solution for creating 0, and try to change it into a solution for the number 1.

» *How can the number line help you develop new strategies?* It helps you visualize and monitor the sizes of the fractions. Instead of focusing only on rules for getting common denominators and equivalent fractions, you can also think about whether you need larger or smaller fractions in order to reach your "target" value.

» *Why is it useful to keep a record of the things you have tried?* It might save you the trouble of repeating unsuccessful attempts. It may also help you see patterns or make observations that will lead to later solutions.

Solution for #1

Sample response:

$$\boxed{\frac{5}{6}} - \boxed{\frac{2}{4}} - \boxed{\frac{1}{3}} = 0$$

Other possibilities include $\frac{5}{6} - \frac{1}{3} - \frac{2}{4}$, $\frac{1}{3} + \frac{2}{4} - \frac{5}{6}$, and $\frac{2}{4} + \frac{1}{3} - \frac{5}{6}$. (Of course, these will look different on the number line.)

Problem #2

2. Create the number 1.

Questions and Conversations for #2

See Questions and Conversations for #1.

Solution for #2

Sample response: Some students may revise their solution to Problem #1 by adding $\frac{2}{4}$ instead subtracting it, thus increasing the sum by 1!

$$\boxed{\frac{5}{6}} - \boxed{\frac{1}{3}} + \boxed{\frac{2}{4}} = 1$$

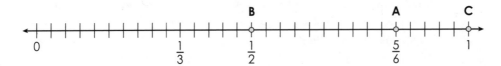

The equation $\frac{2}{4} - \frac{1}{3} + \frac{5}{6} = 1$ also works. (*Note:* The equation $\boxed{\frac{5}{6}} + \boxed{\frac{2}{4}} - \boxed{\frac{1}{3}} = 1$ is also true, but does not fit the conditions of the problem, because it makes B greater than 1.)

Problem #3

3. Create a number as close as possible to but not equal to $\frac{1}{2}$.

Questions and Conversations for #3

See Questions and Conversations for #1. If students have worked for some time without success, consider asking additional questions:

» *What happens to the value of a fraction when you leave its numerator the same and increase its denominator?* Its value decreases.

» *What is the largest denominator you can make for a simplified fraction in this problem?* 60. Students may explain it in many ways. Ultimately, it comes down to the fact that the largest LCM (least common multiple) of any three of the numbers 1, 2, 3, 4, 5, and 6 is 60.

» *Will every fraction that you create fall on the marks of the number line?* Not necessarily. The number line is marked in thirtieths, but you can create sixtieths.

Solution for #3

Sample response:

$$\boxed{\frac{1}{6}} + \boxed{\frac{3}{4}} - \boxed{\frac{2}{5}} = \boxed{\frac{31}{60}}$$

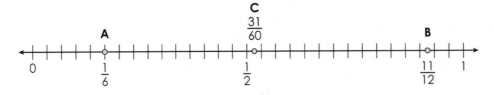

Other solutions: $\dfrac{3}{4} - \dfrac{2}{5} + \dfrac{1}{6} = \dfrac{31}{60}$, $\dfrac{1}{4} + \dfrac{3}{5} - \dfrac{2}{6} = \dfrac{31}{60}$, $\dfrac{3}{5} - \dfrac{2}{6} + \dfrac{1}{4} = \dfrac{31}{60}$

Problem #4

4. Create a number as close as possible to but not equal to 0.

Questions and Conversations for #4

See Questions and Conversations for #1 and #3.

Solution for #4

If students found the solution $\dfrac{3}{5} - \dfrac{2}{6} + \dfrac{1}{4} = \dfrac{31}{60}$ to the previous problem, they may recognize that they can reduce the result from $\dfrac{31}{60}$ to $\dfrac{1}{60}$ by subtracting $\dfrac{1}{4}$ instead of adding it.

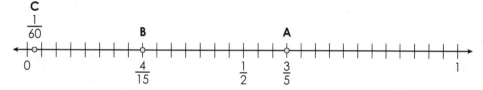

Another solution: $\dfrac{3}{5} - \dfrac{1}{4} - \dfrac{2}{6} = \dfrac{1}{60}$

Problem #5

5. Create a number as close as possible to but not equal to 1.

Questions and Conversations for #5

See Questions and Conversations for #1 and #3.

Solution for #5

Sample response:

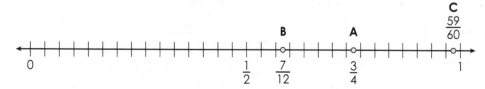

Another response: $\dfrac{2}{5} - \dfrac{1}{6} + \dfrac{3}{4} = \dfrac{59}{60}$

33

STAGE 2

Problem #6

6. Find at least 10 solutions to the equation below. *a*, *b*, and *c* must be whole numbers. Describe your strategies. Show the calculations for at least five of your solutions to prove that they are correct.

$$\frac{1}{a} + \frac{1}{b} + \frac{1}{c} = \frac{1}{5}$$

Teacher's Note. Students should be careful not to confuse *a*, *b*, and *c* in this problem with A, B, and C from the first five problems. They no longer need to check that their intermediate values lie between 0 and 1.

Number lines are not provided for this problem. However, students may benefit from creating their own. It will often be sufficient for them to estimate the positions of points on the line.

Questions and Conversations for #6

» *Can a, b, or c represent the same number?* Yes, any or all them may stand for the same value.

» *Does it count as a new solution if you interchange a, b, and c?* This decision is up to you as the problem solver. However, these additional solutions are easy enough to obtain that most people would not consider them new solutions at all.

» *Is it okay if you get a sum of* $\frac{2}{10}$ *or some other fraction that is equivalent to* $\frac{1}{5}$ *?* Yes. Equivalent fractions represent the same number, so if the sum equals the equivalent fraction, then it equals $\frac{1}{5}$. Consider choosing an equivalent fraction as a starting point for your search.

Teacher's Note. The solutions that students often find first are 15, 15, 15; 10, 10, 20; and 10, 15, 30. This may lead them to conjecture that all three denominators must be multiples of 5. If they test this idea, they should find that at least one denominator, but not necessarily all three, must be a multiple of 5. Can the students see why?

» *Are there any whole numbers that can never be used? Why?* The numbers 1, 2, 3, 4, and 5 cannot be used as denominators because they would create fractions that are greater than or equal to $\frac{1}{5}$. 0 cannot be used because $\frac{1}{0}$ does not represent a number.

» *Must all three denominators be multiples of 5? Must any of them be?* Experiment with other sums of unit fractions (fractions with a numerator of 1) to test this idea.

» *What is the largest fraction that could be one of the addends? How can you find out how much more is needed to make $\frac{1}{5}$? The largest unit fraction smaller than $\frac{1}{5}$ is $\frac{1}{6}$. You can subtract to find out how much you must add to $\frac{1}{6}$ in order to reach $\frac{1}{5}$. Now, how can you use this difference to write a sum of two unit fractions?*

Solution for #6

See the solutions to Problem #7 where all possible answers are listed. Check that students' solutions are somewhere in this list, and they have not repeated any of them.

The strategy that most students employ is to choose a denominator—the number 20, for example—and use this to form a fraction equivalent to $\frac{1}{5}$ ($\frac{4}{20}$ in this case). Then search for three factors of the denominator that add to the numerator. For example, because 1 and 2 are factors of 4, and $4 = 2 + 1 + 1$, they generate the solution $\frac{1}{5} = \frac{4}{20} = \frac{2}{20} + \frac{1}{20} + \frac{1}{20} = \frac{1}{10} + \frac{1}{20} + \frac{1}{20}$. The solutions produced by this strategy are usually the easiest ones to find. However, students may discover that they get a lot of "repeat" solutions when they use this technique.

The clearest way for students to prove that their results are correct is to show the full process for adding the fractions. Here are five sample calculations.

$$\frac{1}{10} + \frac{1}{15} + \frac{1}{30} = \frac{3}{30} + \frac{2}{30} + \frac{1}{30} = \frac{6}{30} = \frac{1}{5}$$

$$\frac{1}{6} + \frac{1}{40} + \frac{1}{120} = \frac{20}{120} + \frac{3}{120} + \frac{1}{120} = \frac{24}{120} = \frac{1}{5}$$

$$\frac{1}{7} + \frac{1}{21} + \frac{1}{105} = \frac{15}{105} + \frac{5}{105} + \frac{1}{105} = \frac{21}{105} = \frac{1}{5}$$

$$\frac{1}{8} + \frac{1}{16} + \frac{1}{80} = \frac{10}{80} + \frac{5}{80} + \frac{1}{80} = \frac{16}{80} = \frac{1}{5}$$

$$\frac{1}{9} + \frac{1}{12} + \frac{1}{180} = \frac{20}{180} + \frac{15}{180} + \frac{1}{180} = \frac{36}{180} = \frac{1}{5}$$

STAGE 3

Problem #7

7. Find all solutions to the equation in Problem #6. Explain how you can be sure you have found them all. (*a*, *b*, and *c* must still be whole numbers.)

Teacher's Note. This may be a challenging approach for students. Most of them do not develop it entirely on their own. It may help to visualize the approximate positions of the sums (like A, B, and C in the Stage 1 questions) on a number line.

Questions and Conversations for #7

> » *Can you think of a way to produce solutions in a systematic way so that you are sure not to miss any?* Find a way to organize them based on the sizes of the fractions. Keep asking: What is the largest fraction I can use? What is the next largest?

Solution for #7

There are 36 solutions to the equation (assuming that changing the order of the values does not count as a new solution). They are listed below as triples: *a, b, c.*

6, 31, 930 *	6, 39, 130 *	6, 55, 66 *	7, 35, 35 +	9, 12, 180 *	10, 15, 30 +
6, 32, 480 *	6, 40, 120	6, 60, 60 +	8, 14, 280 *	9, 15, 45 +	10, 20, 20 +
6, 33, 330 *	6, 42, 105 *	7, 18, 630 *	8, 15, 120	9, 18, 30	11, 11, 55 +
6, 34, 255 *	6, 45, 90	7, 20, 140	8, 16, 80	10, 11, 110	12, 12, 30 +
6, 35, 210 *	6, 48, 80 *	7, 21, 105	8, 20, 40 +	10, 12, 60 +	12, 15, 20
6, 36, 180	6, 50, 75	7, 30, 42	8, 24, 30	10, 14, 35	15, 15, 15 +

The solutions that students find most frequently (in my experience) are marked with a plus symbol. Those least often found are marked with an asterisk.

To find all solutions, be systematic in your search. For example, because all denominators must be greater than 5, begin with 6. Because $\frac{1}{5} - \frac{1}{6} = \frac{1}{30}$, if the first fraction is $\frac{1}{6}$, then the sum of the final two fractions must be $\frac{1}{30}$. Therefore, these two denominators must both be greater than 30. Begin with a denominator of 31. Because $\frac{1}{30} - \frac{1}{31} = \frac{1}{930}$, you can see that $\frac{1}{5} = \frac{1}{6} + \frac{1}{30} = \frac{1}{6} + \frac{1}{31} + \frac{1}{930}$. Continue by testing 32, 33, etc., trying to create sums of $\frac{1}{30}$, checking to see that both fractions have a numerator of 1. When you reach 6, 60, 60, you are finished with denominators of 6, because if you try a denominator greater than 60 in the second fraction, then the denominator of the third fraction will have to be less than 60, and you have already tested all of these possibilities.

The next step is to continue with denominators of 7: $\frac{1}{5} - \frac{1}{7} = \frac{2}{35}$. Because $\frac{1}{18}$ is the largest unit fraction that is less than $\frac{2}{35}$, you can begin with it. Now, because $\dfrac{2}{35} - \dfrac{1}{18} = \dfrac{1}{630}$, you obtain the result $\dfrac{1}{5} = \dfrac{1}{7} + \dfrac{2}{35} = \dfrac{1}{7} + \dfrac{1}{18} + \dfrac{1}{630}$. You can continue testing each whole number denominator greater than 18 until you reach $\dfrac{1}{5} = \dfrac{1}{7} + \dfrac{1}{35} + \dfrac{1}{35}$.

Continue in this vein, testing the whole numbers 6 through 15 as initial denominators. When you reach 15, you are finished, because if you begin with $\dfrac{1}{16}$, you need a value greater than this for at least one of the remaining fractions. This requires a denominator smaller than 16, and you have already tested all of these possibilities.

WRAP UP

Share Strategies

Have students share solutions and strategies. Ask them to compare their solutions to the Stage 1 problems. Draw special attention to any methods that are based on estimation or thinking about the sizes of the fractions.

Students may also notice some interesting patterns involving addition and subtraction, especially if you now remove the restriction that A, B, and C must all be between 0 and 1. For example, the solution, $\frac{2}{4}-\frac{1}{3}+\frac{5}{6}=1$, to Problem #2 can then be written in five other ways using the same three fractions:

$$\frac{2}{4}+\frac{5}{6}-\frac{1}{3} \qquad \frac{5}{6}+\frac{2}{4}-\frac{1}{3} \qquad \frac{5}{6}-\frac{1}{3}+\frac{2}{4} \qquad -\frac{1}{3}+\frac{5}{6}+\frac{2}{4} \qquad -\frac{1}{3}+\frac{2}{4}+\frac{5}{6}$$

Ask students to compare and contrast all six expressions and explain why it makes sense that all of them have the same value, 1.

Summarize

Answer any remaining questions that students have. For the Stage 2 question, summarize and extend the ideas by asking students:

» How many strategies did you find? What were the advantages and disadvantages of each?

» Why can't 0 be used as a denominator? (This may lead to a discussion of why it makes no sense to divide by 0.)

» Why must at least one fraction have a denominator of 5? (Students might think about what happens when they find the least common multiple of all three denominators.)

Compile solutions found by students. Are there any that no one was able to find? If so, you might want to share these with students. Discuss how to use a systematic approach to find all of them.

Further Exploration

Ask students to think of new questions to ask or ways to extend this exploration. Here are some possibilities:

» For Problems #1–#5, how many different fractions can you make? Is it possible to make all of the sixtieths between 0 and 1?

» What happens if you allow the number line in Problems #1–#5 to run between -1 and 1?

» Suppose you change the equation in Problem #6 so that all of the numerators are 2 (or some other number) instead of 1. How does this affect the solutions?

» Can you write $\frac{1}{5}$ as a sum of four, five, or more unit fractions? Are there any patterns in the process of finding these sums? Is there a limit to how many unit fractions you can add to produce $\frac{1}{5}$?

» Is it possible to write every fraction as a sum of unit fractions whose denominators are all different?

» Is there a way to predict the number of solutions to this type of problem?

Exploration 3

Working Together

INTRODUCTION

Prior Knowledge

- » Know procedures for adding fractions.
- » Have experience using variables to represent real-world quantities.

Learning Goals

- » Use pictorial models to represent fraction addition.
- » Explore connections between fractions and division.
- » Solve problems that require working between two different units.
- » Build a foundation for understanding *reciprocals*.
- » Make and test mathematical conjectures.
- » Use algebraic expressions to represent and analyze patterns in computations.
- » Communicate complex mathematical ideas clearly.
- » Persist in solving challenging problems.

Launching the Exploration

Motivation and purpose. To students: Problems like the ones in this exploration often appear in beginning algebra texts. However, you can solve them before you learn traditional algebra procedures. By using prealgebra knowledge and strategies instead, you may actually learn more about the problems than many algebra students do. In particular, you will gain problem-solving experience and a deeper understanding of fractions and unit rates.

Understanding the problem. Read the Motivation and Purpose paragraph to students. Then read the problem statement with them and look through the exploration to help them see the big picture. Problem #2 is important. Although there are strategies that lead to quick solutions, many students will spend a substantial amount of time on this question. The main goal right now is not efficiency but understanding. Each student will learn the most by developing his or her own approach. Once Problem #2 is complete, the task is to analyze the solution, look for patterns and shortcuts, test them, and analyze the results.

As students begin to work, use the first two or three questions from the "Questions and Conversations" section to generate discussion about some important issues and to ensure that they understand the problem.

STUDENT HANDOUT

Stage 1

Hannah can paint a large wall in 3 hours. Glen can do it in 5 hours. If they work together, how long will it take them? Assume that each person works at a constant rate.

The problems below will guide you through the process of exploring this problem.

1. Draw a diagram of the wall showing the fraction that each person paints in one hour. Label each person's contribution clearly. What fraction of the wall have they painted together after one hour? Use your diagram to explain.

2. Use your diagram to determine how long it will take Hannah and Glen to paint the entire wall together. Explain your strategies in detail.

3. Compare the original numbers (3 and 5) to the solution written in hours as an improper fraction. (Rewrite it in this form if necessary.) Predict a quick method to get this answer from the original numbers.

Now suppose that Hannah can paint the wall in 2 hours and Glen can do it in 7 hours.

4. Show how to use your quick method to predict how long it will take them if they work together.

5. Now show how to find this answer with a strategy like the one you used in Problem #2. Write your result in hours as an improper fraction, and compare it your prediction in Problem #4. Do your answers agree?

6. Explain why your quick method in Problem #3 works.

7. Suppose that Hannah and Glen work at the same rate. How long will it take them if they work together? (Use your common sense to answer this question as quickly and easily as you can, but be sure to explain your thinking.)

8. Now use your method from Problem #3 to answer Problem #7. Do your answers agree?

Stage 3

9. Trudi, Ursula, and Verity are working together to complete a job. (You may decide what the task is this time!) Each person works at a constant rate. When Trudi and Ursula work together, they can complete the job in 4 days. Trudi and Verity can finish the task in 3 days. Ursula and Verity can do it 2 days. How long would it take each person to finish the task alone?

TEACHER'S GUIDE

STAGE 1

Hannah can paint a large wall in 3 hours. Glen can do it in 5 hours. If they work together, how long will it take them? Assume that each person works at a constant rate.

Problem #1

1. Draw a diagram of the wall showing the fraction that each person paints in one hour. Label each person's contribution clearly. What fraction of the wall have they painted together after one hour? Use your diagram to explain.

Questions and Conversations for #1

This section contains ideas for conversations, mainly in the form of questions that students may ask or that you may pose to them. Be sure to allow students to do most of the thinking and talking!

» *Why is the answer not just the average of the two times?* The average is 4 hours. It does not make sense that it would take them longer working together than it would take Hannah by herself.

» *What fraction of the wall can Hannah paint in one hour? What fraction can Glen paint?* (These are Hannah's and Glen's *unit rates*.) Hannah can paint $\frac{1}{3}$ of the wall in an hour. Glen can paint $\frac{1}{5}$ of it.

» *Into how many parts can you divide the wall to show both contributions clearly?* The simplest option is probably to use 15 parts, because this is a common multiple of 3 and 5.

» *How can you use your diagram to estimate the time it will take them together?* Your diagram should show that they have painted a little more than half of the wall in an hour. Therefore, it will take a little less than 2 hours.

Teacher's Note. Some students choose 60 parts, because they are thinking of the relationship between hours and minutes. This may work for them because 60 is a multiple of 15. However, they may be confusing the units. The diagram represents one wall, not one hour. This method may cause confusion when they reach Problem #4, because 60 will no longer be a multiple of both times.

Solution for #1

Splitting the wall into 15 equal parts makes it easy to show both thirds and fifths. (Students may make other workable choices such as 30 or 60 parts.) Hannah's part is $\frac{1}{3}$ of the wall, or 5 out of 15 sections. Glen's part is $\frac{1}{5}$ of the wall, or 3 out of 15 sections.

The diagram shows that $\frac{8}{15}$ of the wall has been painted after one hour. Students should recognize this as the sum of $\frac{1}{3}$ and $\frac{1}{5}$.

Problem #2

2. Use your diagram to determine how long it will take Hannah and Glen to paint the entire wall together. Explain your strategies in detail.

Questions and Conversations for #2

» *Once you know the amount they can paint in one hour, can you find how much they can do in smaller units of time?* Yes. Consider trying this for simple fractions such as $\frac{1}{2}$ hour.

» *How can a region in your diagram represent more than one fraction?* It depends on what you take as the whole. Consider thinking of the whole wall and a whole hour.

> **Teacher's Note.** Some students may spend quite a while on Problem #2. They will often think in terms of minutes and seconds. Although this is generally not the quickest way to solve the problem, it is a more intuitive approach for many students. It is important for them to begin by developing a method that makes sense to them.

Solution for #2

Sample response 1: There are 7 sections of the wall still to be painted. Because 8 sections of the wall take 1 hour to paint:

    4 sections of the wall take 30 minutes
    2 sections of the wall take 15 minutes
    <u>1 section of the wall takes 7.5 minutes</u>
    7 sections of the wall take (30 + 15 + 7.5) minutes = 52.5 minutes to paint.

Combining this with the 1 hour required to paint the original 8 sections, the total time needed to paint the wall is 1 hour 52.5 minutes or 1 hour 52 minutes 30 seconds.

Sample response 2: If 8 sections take 1 whole hour to paint, then 7 sections must take $\frac{7}{8}$ of an hour. Therefore, the total time is $1\frac{7}{8}$ hours.

Sample response 3: If 8 sections represent 1 hour, then each section represents $\frac{1}{8}$ of an hour. Because there are 15 sections all together, it will take a total of $\frac{15}{8}$, or $1\frac{7}{8}$ hours.

Sample response 4: The number of hours required is equal to the number of times that 8 sections "fit into" the 15 sections, so you can divide 15 by 8 to get 1.875 hours.

Sample response 5: To bring $\frac{8}{15}$ of the whole up to 1 whole, you have to find a value such that $\frac{8}{15} \cdot ? = 1$.

This is equivalent to asking how many groups of $\frac{8}{15}$ are in 1. The entire $\frac{8}{15}$ plus another $\frac{7}{8}$ of it are needed. (Students who know the standard procedure for multiplying fractions may recognize that the fraction $\frac{15}{8}$ will satisfy the equation.)

Teacher's Note. Students may confuse the units (the whole). For example, when they see that their diagram suggests an answer just short of 2 hours, they will notice that they are one "box" short.

This box represents $\frac{1}{15}$ of the wall, but they might interpret it as $\frac{1}{15}$ of an hour.

Because $\frac{1}{15}$ of an hour is 4 minutes, they may subtract this from 2 hours, obtaining an incorrect answer of 1 hour 56 minutes.

Problem #3

3. Compare the original numbers (3 and 5) to the solution written in hours as an improper fraction. (Rewrite it in this form if necessary.) Predict a quick method to get this answer from the original numbers.

Questions and Conversations for #3

» *How can you convert a time expressed in minutes and seconds into a fraction of an hour?* One way to do it is to think of each minute as $\frac{1}{60}$ of an hour (and each second as $\frac{1}{3600}$ of an hour).

> » *What should you watch for when you compare the original numbers to the answer?* Look for simple calculations that you can do with 3 and 5 to get the answer.

> » *Do you need to understand why these calculations work?* Not yet—but keep the question in mind. It will come up again in Stage 2.

Solution for #3

For students who need to rewrite their answer in the form of an improper fraction, the process could look like this:

Because 52.5 minutes is $\dfrac{52.5}{60}$ of an hour, $\dfrac{52.5}{60} = \dfrac{105}{120} = \dfrac{7}{8}$.

This gives a total time of $1\dfrac{7}{8}$, or $\dfrac{15}{8}$ hours.

Sample response 1: Multiply the two original numbers to get the numerator and add them to get the denominator. In this case, $\dfrac{3 \cdot 5}{3+5} = \dfrac{15}{8}$.

Sample response 2: Add the unit rates and then "invert" the result. $\dfrac{1}{3} + \dfrac{1}{5} = \dfrac{8}{15}$.
When you invert $\dfrac{8}{15}$, you get $\dfrac{15}{8}$.

Problem #4

Now suppose that Hannah can paint the wall in 2 hours and Glen can do it in 7 hours.

4. Show how to use your quick method to predict how long it will take them if they work together.

> **Teacher's Note.** We are saving the use of the term *reciprocal* for Exploration 6 when the real meaning of the concept is developed.

Solution for #4

Using the first method, you get $\dfrac{2 \cdot 7}{2+7} = \dfrac{14}{9} = 1\dfrac{5}{9}$ hours. The second method produces the same result.

Problem #5

5. Now show how to find this answer with a strategy like the one you used in Problem #2. Write your result in hours as an improper fraction, and compare it to your prediction in Problem #4. Do your answers agree?

Questions and Conversations for #5

> » *Is it necessary to use exactly the same strategies that you used in Problem #2?* No. You should still draw a picture and think of unit rates. However, you may

discover new ideas as you work. The key is to find an independent way to verify your prediction in Problem #4.

Solution for #5

Hannah's unit rate is $\frac{1}{2}$ of a room per hour. Glen's unit rate is $\frac{1}{7}$ of a room per hour. This diagram shows each person's contribution after one hour.

Hannah has done 7 out of 14 parts ($\frac{1}{2}$ of the work). Glen has done 2 out of 14 parts ($\frac{1}{7}$ of the work). The picture shows that $\frac{9}{14}$ of the wall has been painted.

Because 9 sections take 1 hour to paint, 1 section takes $\frac{1}{9}$ of an hour. Therefore, the remaining 5 sections take $\frac{5}{9}$ of an hour. The total time needed is $1\frac{5}{9}$ hours. This is consistent with the prediction in Problem #4.

Some students' answers will likely be in this form: 1 hour 33 minutes, 20 seconds. To see that this is the same as $1\frac{5}{9}$ hours, they could begin by finding that $\frac{1}{9}$ of an hour is $60 \div 9 = 6\frac{2}{3}$ minutes. $\frac{5}{9}$ of an hour is five times as much:

$$6\frac{2}{3} \cdot 5 = 6 \cdot 5 + \frac{2}{3} \cdot 5 = 30 + \frac{10}{3} = 30 + 3\frac{1}{3} = 33\frac{1}{3} \text{ minutes, or } 33 \text{ minutes } 20$$

seconds.

Notice that students can do these calculations before learning formal procedures for multiplying fractions and mixed numbers.

STAGE 2

Problem #6

6. Explain why your quick method in Problem #3 works.

Questions and Conversations for #6

In the discussions for Problems #6–#8, a and b represent the two numbers in the problem.

For students whose quick method is based on the expression $\frac{a \cdot b}{a + b}$:

» *Where does the product of a and b appear in your diagrams?* It is the number of parts in the entire diagram. (If students used more than 15 parts in their diagram for Problem #1, ask them to consider how they could have drawn it using fewer parts.)

» *Where does the sum of a and b appear in your diagrams?* It represents the total number of parts completed by both people in an hour.

» *Why does it make sense to place the product and sum into a fraction?* Think of the fraction in terms of division. Or focus on using *hours* as the whole.

For students whose quick method is to add the unit rates and "invert" the result:

» *What does the sum of the unit rates represent?* It stands for the total fraction of the wall completed by both people in one hour.

» *Why do you invert this sum?* Think about the diagram in two ways: with the wall as the whole and with an hour as the whole.

Solution for #6

To make the conclusions valid for a range of values for Hannah's and Glen's times, you can use variables such as a and b to represent them. Then Hannah's unit rate is $\frac{1}{a}$ and Glen's unit rate is $\frac{1}{b}$. In order to show the fraction completed by both people, imagine drawing a diagram of the wall divided into $a \cdot b$ equal parts. The total number of these parts that can be completed in one hour when the two work together is $a + b$. To find the time required to complete the job, divide the total number of parts by the number of parts that can be finished per hour: $(a \cdot b) \div (a + b)$. This may be expressed in fraction form as $\frac{a \cdot b}{a + b}$.

Students whose shortcut was to add the unit rates and "invert" the result might start by observing that the sum of the unit rates represents the fraction of the wall completed by both in 1 hour. To find number of hours required to complete the wall, think of the number of hours as the new whole. To illustrate:

If the entire rectangle (1 wall) is the whole, then it contains 15 equal parts. The shaded region (1 hour) contains 8 of them, so this region is $\frac{8}{15}$ of the whole. If the shaded region (1 hour) is the whole, then it contains 8 equal parts. The entire rectangle (1 wall) contains 15 of these, so it represents $\frac{15}{8}$ of the whole. When you change the whole, you interchange the numbers in the numerator and denominator!

Teacher's Note. Some students may be able to justify these conclusions using more formal algebraic methods. The portion completed by Hannah in 1 hour is $\frac{1}{a} = \frac{b}{a \cdot b}$. The corresponding fraction for Glen is $\frac{1}{b} = \frac{a}{a \cdot b}$. The sum of these is $\frac{a+b}{a \cdot b}$. When you invert this, you get $\frac{a \cdot b}{a+b}$.

Problem #7

7. Suppose that Hannah and Glen work at the same rate. How long will it take them if they work together? (Use your common sense to answer this question as quickly and easily as you can, but be sure to explain your thinking.)

Questions and Conversations for #7

» *Will the answer be a number?* No, because we do not have numbers for Hannah's and Glen's times. The answer will be an algebraic expression.

» *How can you use algebra to show that both people work at the same rate?* Use the same variable (*t*, for example) to represent each person's time.

Solution for #7

Because Hannah and Glen are working at the same speed, it should take them half as long to finish it when they work together. If *t* represents the time for each to do the job alone, their time together is $t \div 2$, $\frac{1}{2} \cdot t$, or $\frac{t}{2}$ hours.

Problem #8

8. Now use your method from Problem #3 to answer Problem #7. Do your answers agree?

Questions and Conversations for #8

» *Can you manipulate fractions with variables using the same procedures you use for numbers?* Yes, because when you use variables, you are representing number patterns. For example, you can still add fractions by finding a common denominator and then adding the numerators. And you can still simplify fractions by multiplying or dividing the numerator and denominator by the same (nonzero) quantity.

Solution for #8

If *a* and *b* are both equal to the same value, *t*, then the first shortcut becomes $\frac{a \cdot b}{a+b} = \frac{t \cdot t}{t+t} = \frac{t \cdot t}{2 \cdot t}$.

If you divide the numerator and the denominator by *t* in the final fraction, it "undoes" the multiplication, leaving $\frac{t}{2}$.

If students' quick method was to add the unit rates and the invert the result, then:

$$\frac{1}{a}+\frac{1}{b}=\frac{1}{t}+\frac{1}{t}=\frac{2}{t}.$$

When you invert the second fraction, you get $\frac{t}{2}$, the same result!

STAGE 3

Problem #9

9. Trudi, Ursula, and Verity are working together to complete a job. (You may decide what the task is this time!) Each person works at a constant rate. When Trudi and Ursula work together, they can complete the job in 4 days. Trudi and Verity can finish the task in 3 days. Ursula and Verity can do it 2 days. How long would it take each person to finish the task alone?

Questions and Conversations for #9

» *Can you apply the same strategy for this problem that you used in Problem #1?* Possibly not. The question is reversed. You are given the amount completed by people working together and asked to find the individual times.

» *What happens if you make the number of days the same for each pair?* It is possible to choose the number of days in a way that makes the problem easier. Think about how many times they could complete the task in this number of days.

Solution for #9

Sample response: Consider making the number of days constant, and find out how many times each pair can do the job in that time frame. For example, in 24 days:

Trudi and Ursula can do the job 6 times.

Trudi and Verity can do the job 8 times.

Ursula and Verity can do the job 12 times.

If T, U, and V represent the number of times that Trudi, Ursula, and Verity, respectively, can complete the job in 24 days, you have the relationships:

$$T+U=6$$
$$T+V=8$$
$$U+V=12$$

Use trial and error or other methods to see that $T=1$, $U=5$, and $V=7$. Therefore:

Trudi can complete the job 1 time in 24 days.

Ursula can complete the job 5 times in 24 days or once in $\frac{24}{5}=4\frac{4}{5}$ days.

Verity can complete the job 7 times in 24 days or once in $\frac{24}{7}=3\frac{3}{7}$ days.

WRAP UP

Share Strategies

Allow students to share and critique their various strategies and observations. There are likely to be many of them!

Summarize

Answer any remaining questions that students have. Summarize key ideas:

» A *unit rate* is a rate in which one of the quantities in the comparison has a value of *one*. In this activity, it is the fraction of the wall painted in *one* hour.

» Problems involving fractions may have more than one unit or whole. In this case, it is important to be clear about the whole to which each fraction refers.

» Discuss the reasons that the "quick methods" developed by students work (see the "Solutions" to Problem #6). Especially talk about why the fraction of the job completed in one hour was "flipped" to obtain the time required to complete one job.

Further Exploration

Ask students to think of new questions to ask or ways to extend this exploration. Here are some possibilities:

» Graph the relationship between the time and fraction of the work completed for each person and for both together. Do the graphs help you understand the problem in new ways?

» Modify the conditions of the original problem to create new ones. For example, consider three or more people working together. Or suppose that you are given a relationship between the rates. (For instance, one person may work twice as fast as the other or take an hour longer than another.)

» What other types of tasks might be involved in problems like the ones in this exploration? What if it is not people but something else doing the work? What if one of the people or things doing the tasks slows the process down?

» Apply your method for solving Problem #9 to the problems at the beginning of the exploration (if you used different methods).

» Let x, y, and z represent the times for the different pairs (T, U; T, V; and U, V respectively) in Problem #9. Find algebraic expressions that would produce the three solutions.

Note. This problem may be too hard for students who have not learned formal algebra procedures, but do not underestimate their ingenuity and persistence! These students often make observations that will not occur to those of us who have come to rely on standard algebraic techniques.

There are many sets of expressions that work. Below are two examples. Multiplication symbols are not explicitly shown.

Trudi: $\dfrac{2}{\dfrac{1}{y}+\dfrac{1}{z}-\dfrac{1}{x}}$ Ursula: $\dfrac{2}{\dfrac{1}{x}+\dfrac{1}{z}-\dfrac{1}{y}}$ Verity: $\dfrac{2}{\dfrac{1}{x}+\dfrac{1}{y}-\dfrac{1}{z}}$

Trudi: $\dfrac{2xyz}{xz+xy-yz}$ Ursula: $\dfrac{2xyz}{yz+xy-xz}$ Verity: $\dfrac{2xyz}{yz+xz-xy}$

Exploration 4

Fractions Forever!

INTRODUCTION

Materials

- » Calculators
- » Colored pencils, crayons, or markers
- » Graph paper (optional)

Prior Knowledge

- » Be fluent with procedures for adding fractions.
- » Use division to convert numbers from fraction to decimal form.
- » Write basic fractions as repeating decimals.

Learning Goals

- » Represent sums of fractions using area models.
- » Describe, analyze, and extend patterns in fraction computations.
- » Recognize that the value of a number is independent of its representation.
- » Use area models to explore *limits* and *geometric series* intuitively.
- » Recognize and represent repeating decimals as infinite sums (Stage 3).
- » Create and justify strategies for writing repeating decimals as fractions (Stage 3).
- » Communicate complex mathematical ideas clearly.
- » Persevere in solving challenging problems.

Launching the Exploration

Motivation and purpose. To students: Is it possible to add an infinite number of numbers? The definition of addition that you learned when you were younger certainly does not cover this situation! On the other hand, imagine that a sum like $0.3 + 0.03 + 0.003 + 0.0003$ continues forever according to the same pattern. Should this equal $0.\overline{3}$—or not? In this exploration, you will use geometric designs to find answers to questions like this.

Understanding the problem. Look through the exploration with students. They will spend most of their time creating and analyzing visual models of fractions to find "infinite sums." Toward the end, they will apply what they have learned to write repeating decimals in fraction form.

Ask the students to describe their observations of the drawing in Problem #1. What would happen if they continued the pattern of filling in the small white square in the upper left part of the picture? Does the drawing help them to imagine what it could mean to "add an infinite number of numbers"? They will see that there is a limit to how large the answers will get as they keep adding more and more numbers.

Teacher's Note. I use the phrase "infinite sum" in this exploration to connect the concept to students' experience and intuition. However, I place it in quotation marks to remind them that this is not the usual type of sum. The formal term for this concept is *infinite series*.

STUDENT HANDOUT

Stage 1

1. The large square has an area of 1 unit². Find the areas of the eight largest regions inside of it. Describe any patterns that you see.

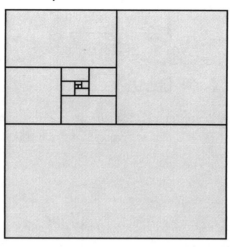

2. Use the drawing to write a list of six equations, beginning with $\frac{1}{2} + \frac{1}{4} = \frac{3}{4}$ and $\frac{1}{2} + \frac{1}{4} + \frac{1}{8} = \frac{7}{8}$. Describe any patterns that make your work easier.

3. How does the drawing illustrate this equation? What do you think the three dots mean?

$$\frac{1}{2} + \frac{1}{4} + \frac{1}{8} + \frac{1}{16} + \frac{1}{32} + \frac{1}{64} + \ldots = 1$$

4. How many of the fractions must you add to get a sum greater than 0.9999? Explain how you found your answer.

5. How close can you get to a sum of 1 by adding more and more fractions? Explain.

Advanced Common Core Math Explorations: Fractions © Prufrock Press Inc.
Permission is granted to photocopy or reproduce this page for single classroom use only.

Exploration 4: Fractions Forever!

6. Draw a picture of a square divided into four congruent parts like the one below. Make it large enough to fill nearly a page. Color the lower left square red, the upper left square yellow, and the lower right square blue. Leave the upper right corner uncolored.

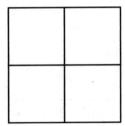

   Now split the uncolored square into four congruent squares, just like the original. Color these four squares in the same way as before. Continue this process of coloring in the blank squares until the squares become too small to show.

7. Use one color to make a list of six equations like those in Problem #2. Explain your thinking, and describe any patterns that you see.

8. Focus on one color to write an equation similar to the one in Problem #3. Write the answer on the right side of the equation as a fraction. Explain how your picture shows this fraction.

9. How can you use decimals to provide evidence that your equation in Problem #8 is true? Show specific calculations to support your answer.

10. Carry out the same type of investigation as in Problems #6–#9 for an "infinite sum" that begins with $\frac{1}{3}$. Consider beginning with this picture:

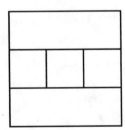

11. Predict the answer to the following based on patterns from equations in Problems #3, #8, and #10. Write your answer in fraction form. Explain your thinking.

$$\frac{1}{10} + \frac{1}{100} + \frac{1}{1000} + \frac{1}{10000} + \frac{1}{100000} + \cdots = ?$$

12. Rewrite this entire equation (including the answer) using only decimals. Does this support your answer to Problem #11? Explain.

You can use place value to write a decimal as a sum of fractions; for example,

$$0.0781 = \frac{7}{100} + \frac{8}{1000} + \frac{1}{10000}.$$

13. Follow this approach to write each repeating decimal as an "infinite sum" of fractions. Then apply your knowledge from this exploration (along with other thinking strategies) to find the value of each decimal as a fraction. Explain your thinking.

a. $0.0\overline{1}$

b. $0.0\overline{4}$

c. $0.2\overline{4}$

d. $0.\overline{36}$

Advanced Common Core Math Explorations: Fractions © Prufrock Press Inc.

TEACHER'S GUIDE

STAGE 1

Problem #1

1. The large square has an area of 1 unit². Find the areas of the eight largest regions inside of it. Describe any patterns that you see.

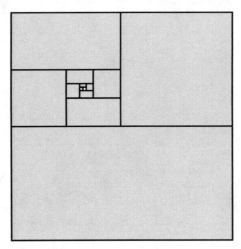

Questions and Conversations for #1

This section contains ideas for conversations, mainly in the form of questions that students may ask or that you may pose to them. Be sure to allow students to do most of the thinking and talking!

» *Where should you focus as you look for the patterns?* Pay attention to the numerators, the denominators, and the overall sizes of the fractions.

» *Is it important to label each area using units²?* Remember that the areas are measured in square units, but you do not have to include the labels in this activity. The focus is on the numbers.

Solution for #1

The eight largest regions have areas of $\frac{1}{2}, \frac{1}{4}, \frac{1}{8}, \frac{1}{16}, \frac{1}{32}, \frac{1}{64}, \frac{1}{128},$ and $\frac{1}{256}$.

The numerator is always 1. The denominators double every time, making each of them a power of 2. Each fraction is half the previous one.

Problem #2

2. Use the drawing to write a list of six equations, beginning with $\dfrac{1}{2} + \dfrac{1}{4} = \dfrac{3}{4}$ and $\dfrac{1}{2} + \dfrac{1}{4} + \dfrac{1}{8} = \dfrac{7}{8}$. Describe any patterns that make your work easier.

Questions and Conversations for #2

» *Do the two given equations count toward the total of six?* Yes.

» *What clues do you have about how the third equation should look?* It should be based on the picture. Look at the two given equations, and figure out how they are related to the picture. Then try to continue the pattern.

Solution for #2

The first six equations are

$$\frac{1}{2} + \frac{1}{4} = \frac{3}{4}$$

$$\frac{1}{2} + \frac{1}{4} + \frac{1}{8} = \frac{7}{8}$$

$$\frac{1}{2} + \frac{1}{4} + \frac{1}{8} + \frac{1}{16} = \frac{15}{16}$$

$$\frac{1}{2} + \frac{1}{4} + \frac{1}{8} + \frac{1}{16} + \frac{1}{32} = \frac{31}{32}$$

$$\frac{1}{2} + \frac{1}{4} + \frac{1}{8} + \frac{1}{16} + \frac{1}{32} + \frac{1}{64} = \frac{63}{64}$$

$$\frac{1}{2} + \frac{1}{4} + \frac{1}{8} + \frac{1}{16} + \frac{1}{32} + \frac{1}{64} + \frac{1}{128} = \frac{127}{128}$$

The patterns from Problem #1 make it easier to write these equations. Also, the denominator of the answer is the same as the denominator of the smallest addend, and the numerator is always 1 less than this.

Problem #3

3. How does the drawing illustrate this equation? What do you think the three dots mean?

$$\frac{1}{2} + \frac{1}{4} + \frac{1}{8} + \frac{1}{16} + \frac{1}{32} + \frac{1}{64} + \ldots = 1$$

Questions and Conversations for #3

» *Why can't you write out every fraction on the left side of the equation?* Because there is no end to them!

» *What does the number 1 on the right side of the equation represent?* Look back to the first question.

Solution for #3

The picture shows rectangles whose areas keep getting half as large. The sum of the areas of all of these rectangles keeps getting closer to 1, the total area of the large square. The three dots indicate that the fractions continue forever in the same pattern.

Problem #4

4. How many of the fractions must you add to get a sum greater than 0.9999? Explain how you found your answer.

Questions and Conversations for #4

» *How can you compare numbers in fraction form to the decimal 0.9999?* Use what you know about connections between fractions and division.

» *Is it okay to use a calculator?* Yes.

Solution for #4

You would need to add 14 of the fractions, up through $\frac{1}{2^{14}}$, or $\frac{1}{16,384}$. The sum of these fractions is $\frac{16,383}{16,384}$, or approximately 0.999939.

Problem #5

5. How close can you get to a sum of 1 by adding more and more fractions? Explain.

Questions and Conversations for #5

» *Can you get within one-millionth of 1? One-billionth? If not, why not? If so, how?* Yes. You can get within one-millionth or one-billionth by adding enough fractions. (If you're not convinced, try it! See how many fractions it takes.)

» *Will you ever reach or surpass the number 1? How do you know?* No, you will never reach or surpass the number 1. The picture provides evidence of this. You will never fill the entire square, because you always fill just a fraction what remains. The pattern in the answers also supports this conclusion because the numerator is always less than the denominator.

Teacher's Note. Students often enjoy acting out a version of this idea based on length rather than area. Ask a volunteer to face a wall at some distance from it. Then have her walk halfway toward it. Repeat the process many times, asking at each stage what fraction of the distance she has covered and what fraction remains. When will she reach the wall? (She will never get there because she covers only a fraction of the remaining distance at each stage.)

Solution for #5

By adding a sufficient number of fractions in the pattern, you can get as close to the number 1 as you like. However, you will never reach or surpass 1, no matter how many of the fractions you add.

STAGE 2

Problem #6

6. Draw a picture of a square divided into four congruent parts like the one below. Make it large enough to fill nearly a page. Color the lower left square red, the upper left square yellow, and the lower right square blue. Leave the upper right corner uncolored.

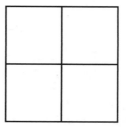

Now split the uncolored square into four congruent squares, just like the original. Color these four squares in the same way as before. Continue this process of coloring in the blank squares until the squares become too small to show.

Solution for #6

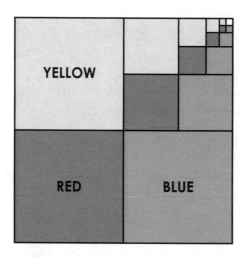

Problem #7

7. Use one color to make a list of six equations like those in Problem #2. Explain your thinking, and describe any patterns that you see.

Solution for #7

$$\frac{1}{4} + \frac{1}{16} = \frac{5}{16}$$

$$\frac{1}{4} + \frac{1}{16} + \frac{1}{64} = \frac{21}{64}$$

$$\frac{1}{4} + \frac{1}{16} + \frac{1}{64} + \frac{1}{256} = \frac{85}{256}$$

$$\frac{1}{4} + \frac{1}{16} + \frac{1}{64} + \frac{1}{256} + \frac{1}{1024} = \frac{341}{1024}$$

$$\frac{1}{4} + \frac{1}{16} + \frac{1}{64} + \frac{1}{256} + \frac{1}{1024} + \frac{1}{4096} = \frac{1365}{4096}$$

$$\frac{1}{4} + \frac{1}{16} + \frac{1}{64} + \frac{1}{256} + \frac{1}{1024} + \frac{1}{4096} + \frac{1}{16,384} = \frac{5461}{16,384}$$

Some students may see patterns in the addends. For example, the fractions keep getting one fourth as large, the numerators are all 1, and the denominators are successive powers of four. The denominator of the answer equals the denominator of the smallest addend. There are some amazing patterns in the numerators! Each numerator is:

  » one more than four times the preceding numerator,
  » one-third of the number that is one less than the denominator,
  » the sum of the previous numerator and denominator, and
  » one more than the sum of all of the denominators except the one in the smallest addend.

Problem #8

8. Focus on one color to write an equation similar to the one in Problem #3. Write the answer on the right side of the equation as a fraction. Explain how your picture shows this fraction.

Questions and Conversations for #8

  » *How do the areas of the red, yellow, and blue parts of your drawing compare?* They are equal! What does this tell you about the size of each region? What does it tell you about your equation?

Solution for #8

$$\frac{1}{4} + \frac{1}{16} + \frac{1}{64} + \frac{1}{256} + \frac{1}{1024} + \cdots = \frac{1}{3}$$

Each color covers an area of $\frac{1}{3}$, because the large square of area 1 has been divided into three regions of equal area.

Problem #9

9. How can you use decimals to provide evidence that your equation in Problem #8 is true? Show specific calculations to support your answer.

Questions and Conversations for #9

» *What should happen as you add more and more numbers?* The answer should get closer and closer to the decimal form of the answer you found in Problem #8.

Solution for #9

If you write the equations in decimal form, it looks like this:

$$\frac{1}{4}+\frac{1}{16}=0.25+0.0625=0.3125$$

$$\frac{1}{4}+\frac{1}{16}+\frac{1}{64}=0.25+0.0625+0.015625=0.328125$$

$$\frac{1}{4}+\frac{1}{16}+\frac{1}{64}+\frac{1}{256}=0.25+0.0625+0.015625+0.00390625=0.33203125$$

As you add more fractions, the answer keeps getting closer to $0.\overline{3}$.

Problem #10

10. Carry out the same type of investigation as in Problems #6–#9 for an "infinite sum" that begins with $\frac{1}{3}$. Consider beginning with this picture:

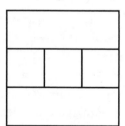

Questions and Conversations for #10

» *Are there other ways to begin the picture?* Yes. In fact, there are many ways to draw all of the pictures in this exploration. Feel free to try some different ones! In order to make them easier to interpret, draw them so that regions of the same color are connected.

Solution for #10

Sample drawings:

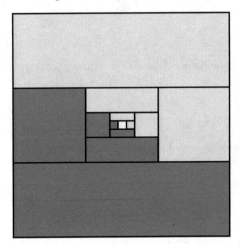 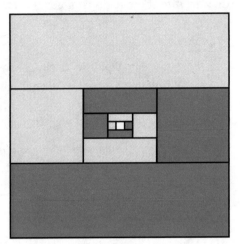

$$\frac{1}{3}+\frac{1}{9}=\frac{4}{9}$$

$$\frac{1}{3}+\frac{1}{9}+\frac{1}{27}=\frac{13}{27}$$

$$\frac{1}{3}+\frac{1}{9}+\frac{1}{27}+\frac{1}{81}=\frac{40}{81}$$

$$\frac{1}{3}+\frac{1}{9}+\frac{1}{27}+\frac{1}{81}+\frac{1}{243}=\frac{121}{243}$$

$$\frac{1}{3}+\frac{1}{9}+\frac{1}{27}+\frac{1}{81}+\frac{1}{243}+\frac{1}{729}=\frac{364}{729}$$

$$\frac{1}{3}+\frac{1}{9}+\frac{1}{27}+\frac{1}{81}+\frac{1}{243}+\frac{1}{729}+\frac{1}{2187}=\frac{1093}{2187}$$

Each addend is one-third as large as the previous one. The denominators are all powers of 3. The denominator of the answer is equal to the denominator of the smallest addend. Each numerator is:

  » one more than three times the preceding numerator,
  » one-half of the number that is one less than the denominator,
  » the sum of the previous numerator and denominator, and
  » one more than the sum of the denominators of all but the smallest addend.

Because each color covers a region with half the area of the whole square, you get the equation:

$$\frac{1}{3}+\frac{1}{9}+\frac{1}{27}+\frac{1}{81}+\frac{1}{243}+\frac{1}{729}+\cdots=\frac{1}{2}$$

If you write this in decimal form, the answers get closer and closer to 0.5.

STAGE 3

Problem #11

11. Predict the answer to the following based on patterns from equations in Problems #3, #8, and #10. Write your answer in fraction form. Explain your thinking.

$$\frac{1}{10} + \frac{1}{100} + \frac{1}{1000} + \frac{1}{10,000} + \frac{1}{100,000} + \cdots = ?$$

Questions and Conversations for #11

» *What might help you find patterns in the equations?* Write down the three equations that involve "infinite sums." List them in order of decreasing sums. Compare the fractions on the left side to the final answer.

Solution for #11

Analyze the equations you have explored so far:

$$\frac{1}{2} + \frac{1}{4} + \frac{1}{8} + \frac{1}{16} + \frac{1}{32} + \frac{1}{64} + \frac{1}{128} + \cdots = 1$$

$$\frac{1}{3} + \frac{1}{9} + \frac{1}{27} + \frac{1}{81} + \frac{1}{243} + \frac{1}{729} + \cdots = \frac{1}{2}$$

$$\frac{1}{4} + \frac{1}{16} + \frac{1}{64} + \frac{1}{256} + \frac{1}{1024} + \cdots = \frac{1}{3}$$

In each case, the denominators are powers of the denominator of the first (largest) fraction, and the denominator of the answer is 1 less than this number. Based on this, you would predict that:

$$\frac{1}{10} + \frac{1}{100} + \frac{1}{1000} + \frac{1}{10,000} + \frac{1}{100,000} + \cdots = \frac{1}{9}$$

Problem #12

12. Rewrite this entire equation (including the answer) using only decimals. Does this support your answer to Problem #11? Explain.

Questions and Conversations for #12

» *Does the decimal form of your answer agree with its fraction form in Problem 11? How can you tell?* It should. Use division to write the fraction as a decimal and then compare.

Solution for #12

$$0.1 + 0.01 + 0.001 + 0.0001 + 0.00001 + 0.000001 + \cdots = 0.\overline{1}$$

This supports the previous conclusion, because the decimal form for $\frac{1}{9}$ is $0.\overline{1}$.

Problem #13

You can use place value to write a decimal as a sum of fractions; for example:

$$0.0781 = \frac{7}{100} + \frac{8}{1000} + \frac{1}{10000}$$

13. Follow this approach to write each repeating decimal as an "infinite sum" of fractions. Then apply your knowledge from this exploration (along with other thinking strategies) to find the value of each decimal as a fraction. Explain your thinking.
    a. $0.0\overline{1}$
    b. $0.0\overline{4}$
    c. $0.2\overline{4}$
    d. $0.\overline{36}$

Teacher's Note. The condition that $\frac{1}{10}$ of a number is equal to $\frac{1}{10}$ less than the number actually determines the number $\frac{1}{9}$. Consider the algebraic equation

$$\frac{1}{10} \cdot x = x - \frac{1}{10}.$$

Those who know procedures for solving algebraic equations may verify that the solution is $\frac{1}{9}$! By generalizing this idea to "infinite sums" similar to those in this activity, you can show that $\frac{1}{n} + \frac{1}{n^2} + \frac{1}{n^3} + \frac{1}{n^4} + \cdots = \frac{1}{n-1}$ is true for any natural number n that is greater than 1. (What if $n = 1$? What about other values of n?)

Questions and Conversations for #13

» *How do the terms* in Part A compare with the ones in Problem #11?* Each term is $\frac{1}{10}$ as large as the corresponding one in Problem #11. Some students may observe instead that the first term, $\frac{1}{10}$, is simply missing from the expression in this question.

» *What is the effect in Part B of making each fraction four times as large as in Part A?* It makes the sum four times as large as well.

» *How does the value of the decimal (or sum) in Part C compare to the one in Part D?* It is greater by two-tenths.

*The *terms* are the "addends."

> » *What is the value of* $\dfrac{1}{100} + \dfrac{1}{10{,}000} + \dfrac{1}{1{,}000{,}000} + \cdots$ *? How can you use this in Part D?* Its value is $\dfrac{1}{99}$. Consider the possibility of using two-digit numerators when you write this decimal as a sum of fractions.

Solution for #13

a. $0.0\overline{1} = \dfrac{1}{100} + \dfrac{1}{1000} + \dfrac{1}{10{,}000} + \dfrac{1}{100{,}000} + \cdots = \dfrac{1}{90}$

Sample response 1: Each fraction in this expression is $\dfrac{1}{10}$ as large as the corresponding fraction in the equation from Problem #11. Therefore, the result must also be $\dfrac{1}{10}$ as large, making the answer $\dfrac{1}{90}$ instead of $\dfrac{1}{9}$.

Sample response 2: This expression is the same as the one in Problem #11, except that the fraction $\dfrac{1}{10}$ is missing from the beginning. Therefore the answer is $\dfrac{1}{9} - \dfrac{1}{10} = \dfrac{10}{90} - \dfrac{9}{90} = \dfrac{1}{90}$.

b. $0.0\overline{4} = \dfrac{4}{100} + \dfrac{4}{1000} + \dfrac{4}{10{,}000} + \dfrac{4}{100{,}000} + \cdots = \dfrac{2}{45}$

Because each fraction is four times as large as the corresponding one in the previous expression, the sum should be four times as large: $\dfrac{4}{90} = \dfrac{2}{45}$.

c. $0.2\overline{4} = \dfrac{2}{10} + \dfrac{4}{100} + \dfrac{4}{1000} + \dfrac{4}{10{,}000} + \dfrac{4}{100{,}000} + \cdots = \dfrac{11}{45}$

Because of the 2 in the tenths place, this should be $\dfrac{2}{10}$ (or $\dfrac{1}{5}$) greater than the answer to Part B. $\dfrac{1}{5} + \dfrac{2}{45} = \dfrac{9}{45} + \dfrac{2}{45} = \dfrac{11}{45}$. You can verify this by dividing 11 by 45.

d. Many strategies involve observing that $\dfrac{1}{100} + \dfrac{1}{10{,}000} + \dfrac{1}{1{,}000{,}000} + \cdots = \dfrac{1}{99}$

Sample response 1: $0.\overline{36} = \dfrac{36}{100} + \dfrac{36}{10{,}000} + \dfrac{36}{1{,}000{,}000} + \cdots$ is 36 times larger than the expression above. Therefore, $0.\overline{36} = 36 \cdot \dfrac{1}{99} = \dfrac{36}{99} = \dfrac{4}{11}$.

Sample response 2:

$$0.\overline{36} = 0.\overline{3} + 0.0\overline{3} = \left(\frac{3}{10} + \frac{3}{100} + \frac{3}{1000} + \ldots\right) + \left(\frac{3}{100} + \frac{3}{10,000} + \frac{3}{1,000,000} + \ldots\right)$$

The expression in the first set of parentheses has a value of $\frac{1}{3}$. The expression in the second set of parentheses has a value of $\frac{3}{99} = \frac{1}{33}$, because it has 3 times the value of the expression shown above for $\frac{1}{99}$. Adding these, you get $\frac{1}{3} + \frac{1}{33} = \frac{11}{33} + \frac{1}{33} = \frac{12}{33} = \frac{4}{11}$.

Teacher's Note. We are assuming the validity of the process of regrouping terms in "infinite sums." Students may learn to justify this in future courses.

Sample response 3: You can decompose $0.\overline{36}$ as $0.\overline{30} + 0.\overline{06}$. By applying the same type of thinking to these expressions, you get:

$$0.\overline{30} + 0.\overline{06} = \frac{30}{99} + \frac{6}{99} = \frac{36}{99} = \frac{4}{11}.$$

WRAP UP

Share Strategies

Have students compare and critique one another's drawings, observations, and strategies. Use this opportunity to identify and correct any misconceptions that they have. Discuss the variety of patterns that arise, especially in the numerators of the answers to Problems #3, #8, and #10. If students have discovered different strategies for writing repeating decimals as fractions, talk about the advantages and disadvantages of each.

Summarize

Answer any remaining questions that students have. Summarize and expand on some key ideas:

» It is sometimes possible to define a sum of infinitely many numbers, but it is important that the numbers you are adding get smaller and smaller. Even in this case, there is not always an answer. For example, the expression $1+\dfrac{1}{2}+\dfrac{1}{3}+\dfrac{1}{4}+\dfrac{1}{5}+\dfrac{1}{6}+\cdots$ (known as the *harmonic series*) does not have a sum because—perhaps surprisingly—there is no limit to how large the result will eventually become as you add more and more of the fractions! (Some students may want to explore this expression in more detail.)

» The "infinite sums" we have been working with are actually known as *infinite series*. The individual "addends" are called *terms*. To define *infinite series* carefully, you need concepts from calculus. Repeating decimals represent an example of a special type of infinite series known as a *geometric series*, in which you multiply each term by the same number to obtain the next term.

Further Exploration

Ask students to think of new questions to ask or ways to extend this exploration. Here are some possibilities:

» Create new pictures to illustrate sums you have explored or sums based on other fractions such as $\dfrac{1}{5}$ or $\dfrac{1}{6}$. (The next page contains an example of a method that one student developed for $\dfrac{1}{5}+\dfrac{1}{25}+\dfrac{1}{125}+\dfrac{1}{625}\ldots$.) By the way, you do not always have to start with a square!

» Apply and extend what you have learned in this exploration to find the fraction form of any repeating decimal.

» Learn about "sigma notation" for sums. For example, the sum $\dfrac{1}{2}+\dfrac{1}{4}+\dfrac{1}{8}+\dfrac{1}{16}+\dfrac{1}{32}+\cdots$ may be written as $\displaystyle\sum_{n=1}^{\infty}\dfrac{1}{2^{n}}$.

71

» What happens if you leave out certain terms in your sums? For example, suppose you omit every second fraction in the sum $\frac{1}{2}+\frac{1}{4}+\frac{1}{8}+\frac{1}{16}+\frac{1}{32}+\cdots$ to create the new sum $\frac{1}{2}+\frac{1}{8}+\frac{1}{32}+\frac{1}{128}+\cdots$. Now what is the value? Why? (Answer: $\frac{2}{3}$)

» What happens if the first term does not have the number 1 in the numerator? For example, suppose you begin with $\frac{2}{3}$. (Answer: $\frac{2}{3}+\frac{4}{9}+\frac{8}{27}+\frac{16}{81}+\frac{32}{243}+\cdots=2$. Does this fit the patterns you have discovered so far?)

» Examine this picture produced by a student. Can you see how it works? How large are the squares in the middle? Can you think of a way to use other regular polygons to draw pictures for equations based on $\frac{1}{6}$, $\frac{1}{7}$, etc.?

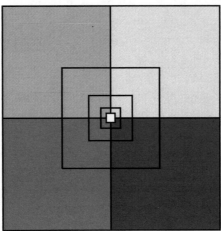

Parker's method for illustrating the equation $\frac{1}{5}+\frac{1}{25}+\frac{1}{125}+\frac{1}{625}+\cdots=\frac{1}{4}$.

Exploration 5

Visualizing Fraction Multiplication

INTRODUCTION

Materials

» Graph paper (recommended)

Prior Knowledge

» Understand addition of fractions.
» Complete Exploration 1: Sharing and Grouping (recommended).
  Note. This activity works best before students learn procedures for multiplying fractions.

Learning Goals

» Use diagrams and stories to illustrate and explain the meaning of fraction multiplication.
» Develop and justify algorithms for multiplying fractions and mixed numbers.
» Discuss advantages and disadvantages of different algorithms for multiplying mixed numbers.
» Understand the changing role of the whole in fraction multiplication.
» Communicate complex mathematical ideas clearly.
» Persist in solving challenging problems.

Launching the Exploration

Motivation and purpose. To students: We often think of multiplication as being harder than addition. In the case of fractions, the *procedure* for multiplication may be easier than you expect, but understanding the meaning is more challenging. The *understanding* piece is your main goal in this exploration.

Understanding the problem. Read the Motivation and Purpose section to students. Then read through the four steps for Problems #1–#3. Make sure students understand what they are being asked to do. (Refer to some examples in the Solutions sections of the Teacher's Guide for this lesson to clarify this for yourself if needed.)

Remind students that a multiplication expression can have two meanings based on the grouping. In order to be as consistent and clear as possible throughout this exploration, let us agree that the first number represents the size of one group, and the second number stands for the number of groups. For example, $3 \cdot 2$ will mean 2 groups of 3, while $2 \cdot 3$ will be 3 groups of 2.

As students begin work, monitor them for a while to offer support as they begin thinking through the process of creating drawings and finding values, but be sure to allow them to develop their own ideas, especially for creating the pictures. Remind them that the goal is to understand the meaning of fraction multiplication, not just a procedure. If some students already know a procedure, challenge them to pretend that no one knows it and that they are trying to be the first to discover one!

Note. Creating stories can be fun for students, but it is not easy and they will be doing a lot of it in this exploration. To help them keep the momentum going throughout the activity, they may enjoy creating an ongoing series of characters, perhaps linking their stories together to create a narrative for them.

STUDENT HANDOUT

Stage 1

For Problems #1–#3:

» Find the value.
» Prove your answer with a drawing or a diagram. Explain your thinking.
» Create a story that fits the situation.
» Write a number model (equation).

1. $\frac{3}{4}$ of a group of 6

2. $\frac{2}{3}$ of a group of $\frac{1}{2}$

3. $\frac{2}{5}$ of a group of $\frac{4}{7}$

4. For one or more of the problems above, follow the same four steps with the numbers reversed: 6 groups of $\frac{3}{4}$, $\frac{1}{2}$ of a group of $\frac{2}{3}$, and $\frac{4}{7}$ of a group of $\frac{2}{5}$.

5. Create an efficient algorithm (a step-by-step procedure) to find the values for Problems #1–#3. Use your pictures to help you explain why your algorithm works.

Advanced Common Core Math Explorations: Fractions © Prufrock Press Inc.
Permission is granted to photocopy or reproduce this page for single classroom use only.

Stage 2

For Problems #6–#8 and #10:

- » Find the value.
- » Prove your answer with a drawing or a diagram. Explain your thinking.
- » Create a story that fits the situation.
- » Write a number model (equation).

6. $\frac{3}{7}$ of a group of 5

7. $3\frac{2}{3}$ groups of 4

8. $2\frac{1}{4}$ groups of $\frac{2}{5}$

9. Does the algorithm that you created in Problem #5 still work for Problems #6–#8? If so, show why. If not, try to modify it or create a new algorithm that will work. Explain your thinking.

Stage 3

10. $1\frac{3}{4}$ groups of $1\frac{1}{2}$

11. If you need to modify an old algorithm or create a new one for Problem #10, do this and explain your thinking. If not, show that one of your earlier algorithms works.

TEACHER'S GUIDE

STAGE 1

For Problems #1–#3:

» Find the value.
» Prove your answer with a drawing or a diagram. Explain your thinking.
» Create a story that fits the situation.
» Write a number model (equation).

Problem #1

1. $\frac{3}{4}$ of a group of 6

Questions and Conversations for #1

This section contains ideas for conversations, mainly in the form of questions that students may ask or that you may pose to them. Be sure to allow students to do most of the thinking and talking! You may easily extend the conversations by creating and exploring additional examples using your own numbers.

» *Can you change the order in which you carry out the four parts of each question?* Yes! Any of the four parts (the value, picture, story, and equation) can support your understanding of the others. Feel free to do them in any order that helps you make the most sense of the expression.

» *What number are you beginning with? What are you doing to this number?* You are beginning with the number 6 and making it smaller by taking part of it.

» *How can you visualize dividing the group of 6 into equal fourths?* Some students might split the group of 6 in half, and then split each of these groups in half. Others may split each "1" into two (or four) equal parts, creating a total of 12 (or 24) parts, which can easily be divided into four equal parts. The second approach is more helpful in most circumstances.

» *What happens to the "whole" as you work?* The whole always begins as the number 1. You

> **Teacher's Note.** Watch for the following misconceptions in students' stories. Many of the errors appear in the ways that they ask their questions. (Correct versions of these questions appear in the Solutions.)
>
> • Students are unclear about the whole to which they are referring. For example: "A bottle holds 6 ounces of glue. You have used $\frac{3}{4}$. How much glue have you used?" There are two wholes involved, 1 ounce and 1 bottle. It is not clear whether the $\frac{3}{4}$ refers to ounces or bottles. This is also somewhat unclear in the question. Are they asking for the number of ounces or bottles used?

Teacher's Note. Continued.

- Students confuse fraction multiplication with subtraction concepts. For example: "A bottle holds 6 ounces of glue. You have used $\frac{3}{4}$ of the bottle for an art project. How much glue is left?" This is a common mistake. Students realize that the answer will be less than 6. Because they are accustomed to making things smaller by using subtraction, they build subtraction ideas into their stories by asking how much is left.

- Students may not refer back to the original whole when they ask their question. For example (see Problem #3): "$\frac{4}{7}$ of the students at Riverview Middle School attended the spring dance. $\frac{2}{5}$ of those who attended stayed to help clean up afterward. How many students helped clean up?" When they ask "How many students . . . ," they fail to recognize that the answer should refer not to a number of students but to a fraction of the students at the entire school.

Teacher's Note. These solutions contain samples of typical responses by students. Their diagrams and stories will almost certainly look different than what is shown here.

start with 6 groups of this, which makes a new whole. You take $\frac{3}{4}$ of this. However, when you find the value at the end of the process, you express it again in terms of the original whole (the number 1).

» *What should the diagram look like?* This is your choice. You may use number lines, rectangles, circles, or anything else that will clearly show the meaning of $\frac{3}{4}$ of a group of 6.

» *Should the story end with a question?* Yes. The question is the part that captures the meaning of the situation.

Solution for #1

Value: $\frac{3}{4}$ of a group of 6 equals $4\frac{1}{2}$, or 4.5.

Sample diagram 1:

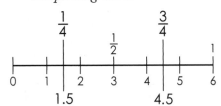

Sample diagram 2:

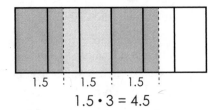

The diagrams show that if you split 6 into 4 equal parts, each part is 1.5. Because $\frac{1}{4}$ of the whole is 1.5, $\frac{3}{4}$ of the whole is $1.5 \cdot 3 = 4.5$.

Sample story: A bottle holds 6 ounces of glue. You have used $\frac{3}{4}$ of the bottle for an art project. How many ounces of glue have you used? (Answer: $4\frac{1}{2}$ ounces of glue)

Number Models: $6 \cdot \frac{3}{4} = 4\frac{1}{2}$ or $6 \div 4 \cdot 3 = 1\frac{1}{2} \cdot 3 = 4\frac{1}{2}$

Problem #2

2. $\frac{2}{3}$ of a group of $\frac{1}{2}$

Questions and Conversations for #2

See Questions and Conversations for #1.

Solution for #2

Value: $\frac{2}{3}$ of a group of $\frac{1}{2}$ equals $\frac{1}{3}$.

Sample diagram 1:   *Sample diagram 2:*

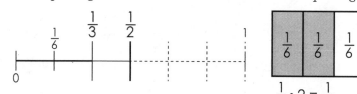

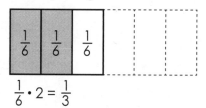

$\frac{1}{6} \cdot 2 = \frac{1}{3}$

The diagrams show that if you split $\frac{1}{2}$ into 3 equal parts, each part is $\frac{1}{6}$. Because $\frac{1}{3}$ of the group of $\frac{1}{2}$ is $\frac{1}{6}$, $\frac{2}{3}$ of it is $\frac{1}{6} \cdot 2 = \frac{2}{6} = \frac{1}{3}$.

Sample story: You and your sister will each mow $\frac{1}{2}$ of the lawn. You have finished $\frac{2}{3}$ of your part. What fraction of the lawn have you mowed? (Answer: $\frac{1}{3}$ of the lawn)

Number Models: $\frac{1}{2} \cdot \frac{2}{3} = \frac{1}{3}$ or $\frac{1}{2} \div 3 \cdot 2 = \frac{1}{6} \cdot 2 = \frac{1}{3}$

Problem #3

3. $\frac{2}{5}$ of a group of $\frac{4}{7}$

Questions and Conversations for #3

See Questions and Conversations for #1.

Solution for #3

Value: $\frac{2}{5}$ of a group of $\frac{4}{7}$ equals $\frac{8}{35}$.

Sample diagram:

Step 1 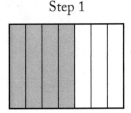   Step 2 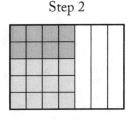   Step 3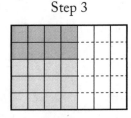

In Step 1, we form a group of $\frac{4}{7}$. In Step 2, we split this group into 5 equal parts and show $\frac{2}{5}$ of it. In Step 3, we show that each small square is $\frac{1}{35}$ of the original whole. Because $\frac{2}{5}$ of the group of $\frac{4}{7}$ holds 8 of these squares, it represents the fraction $\frac{8}{35}$.

Sample story: $\frac{4}{7}$ of the students at Riverview Middle School attended the spring dance. $\frac{2}{5}$ of those who attended stayed to help clean up afterward. What fraction of the students at Riverview Middle School helped clean up? (Answer: $\frac{8}{35}$ of the students)

Number Model: $\frac{4}{7} \cdot \frac{2}{5} = \frac{8}{35}$

Problem #4

4. For one or more of the problems above, follow the same four steps with the numbers reversed: 6 groups of $\frac{3}{4}$, $\frac{1}{2}$ of a group of $\frac{2}{3}$, and $\frac{4}{7}$ of a group of $\frac{2}{5}$.

Questions and Conversations for #4

» *Can't you use the same stories and pictures as in Problems #1–#3?* No, your stories will have to be different. Even if the answers do not change when you reverse the order of the numbers, the meanings do change. You are beginning with different numbers and doing something different to them. The pictures will probably change, too. However, there are some situations in which you might be able to use the same picture and look at it in a different way.

Solution for #4

Students will find that all of the values remain the same when the numbers are reversed. (Multiplication is still *commutative* in these situations.) Here is a sample drawing, story, and number model for the "reversed" version of Problem #1 (6 groups of $\frac{3}{4}$).

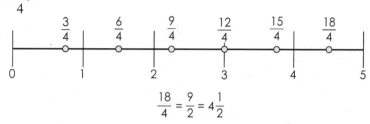

$$\frac{18}{4} = \frac{9}{2} = 4\frac{1}{2}$$

Sample story: If you ride your bike $\frac{3}{4}$ of a mile each day, how many miles have you ridden after 6 days? (Answer: $4\frac{1}{2}$ miles)

Number model: $\frac{3}{4} \cdot 6 = \frac{18}{4} = 4\frac{1}{2}$.

Problem #5

5. Create an efficient algorithm (a step-by-step procedure) to find the values for Problems #1–#3. Use your pictures to help you explain why your algorithm works.

> **Teacher's Note.** Notice how we reversed the order of the numbers $\frac{3}{4}$ and 6 in the number model to be consistent with the decision to let the first number represent the size of a group and the second to stand for the number of groups.

Questions and Conversations for #5

> » *Why aren't there any consistent patterns between the original numbers and the answers in Problems #1–#3?* The patterns may not be obvious if a fraction has been simplified. Problem #3 shows the important pattern most clearly.
>
> » *How does the picture show the denominator?* Every picture shows that the number of parts in one whole is found by multiplying the denominators of the two fractions. (A similar observation holds for the numerator.)

Solution for #5

The results of Problem #3 ($\frac{4}{7} \cdot \frac{2}{5} = \frac{8}{35}$) suggest the algorithm "multiply the numerators and multiply the denominators" or $\frac{a}{b} \cdot \frac{c}{d} = \frac{a \cdot c}{b \cdot d}$.

This makes sense, because $b \cdot d$ represents the total number of parts in the original whole, while $a \cdot c$ represents the number of parts that you "have." (This shows up clearly in the 2 by 4 rectangle in the diagram.)

You can test the algorithm on the results for Problems #1 and #2:

$$6 \cdot \frac{3}{4} = \frac{6}{1} \cdot \frac{3}{4} = \frac{6 \cdot 3}{1 \cdot 4} = \frac{18}{4} = 4\frac{1}{2} \quad \text{and} \quad \frac{1}{2} \cdot \frac{2}{3} = \frac{1 \cdot 2}{2 \cdot 3} = \frac{2}{6} = \frac{1}{3}.$$

Both of them give the same values as before!

STAGE 2

For Problems #6–#8 and #10:

» Find the value.

» Prove your answer with a drawing or a diagram. Explain your thinking.

» Create a story that fits the situation.

» Write a number model (equation).

Problem #6

6. $\frac{3}{7}$ of a group of 5

Questions and Conversations for #6

» *What makes these questions more challenging than Problems #1–#3?* There are two particular features that are different. First, some of the numerators and denominators are chosen in order to make it more challenging to split them into an equal number of parts. Second, the presence of mixed numbers may inspire you to think differently about how to draw your pictures and find values.

» *What kinds of drawings will be most effective for showing the meanings of the expressions?* This may vary from problem to problem and person to person, but try to be flexible. Many students forget about the possibility of using number lines, and they are sometimes easier to draw and understand.

» *When you have a mixed number, can it help to think of the whole number and fractional parts separately?* Yes, you should consider this possibility. It may lead to new strategies for drawing pictures and finding values.

Solution for #6

Value: $\frac{3}{7}$ of a group of 5 equals $2\frac{1}{7}$.

Sample diagram:

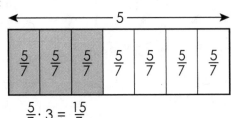

$$\frac{5}{7} \cdot 3 = \frac{15}{7}$$

The diagram shows splitting a group of 5 into 7 equal parts. Each part has a value of $\frac{5}{7}$ (see Exploration 1: Sharing and Grouping). Because $\frac{1}{7}$ of a group of 5 equals $\frac{5}{7}$, $\frac{3}{7}$ of a group of 5 is $\frac{5}{7} \cdot 3 = \frac{15}{7} = 2\frac{1}{7}$.

Sample story: You have written a 5-page story. You completed the first $\frac{3}{7}$ of the story in one night. How many pages of the story did you finish that night? (Answer: $2\frac{1}{7}$ pages)

Number Models: $5 \cdot \frac{3}{7} = \frac{15}{7} = 2\frac{1}{7}$ or $5 \div 7 \cdot 3 = \frac{5}{7} \cdot 3 = \frac{15}{7} = 2\frac{1}{7}$

Students may be interested to notice how this process shows why $5 \cdot \frac{3}{7} = \frac{5}{7} \cdot 3$.

Problem #7

7. $3\frac{2}{3}$ groups of 4

Questions and Conversations for #7

See Questions and Conversations for #6.

Solution for #7

Value: $3\frac{2}{3}$ groups of 4 equals $14\frac{2}{3}$.

Sample diagram:

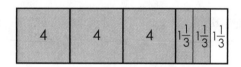

The 3 groups of 4 equal 12, and each third of a group of 4 equals $1\frac{1}{3}$. Therefore, $3\frac{2}{3}$ groups of 4 equals $4 + 4 + 4 + 1\frac{1}{3} + 1\frac{1}{3} = 14\frac{2}{3}$.

Sample story: You are paid \$4 per hour to clean the garage, and you work for $3\frac{2}{3}$ hours. How much do you earn? (Answer: $14\frac{2}{3}$, which represents about \$14.67.)

Number model: $4 \cdot 3\frac{2}{3} = 14\frac{2}{3}$ or $4 \cdot 3 + 4 \cdot \frac{2}{3} = 12 + 2\frac{2}{3} = 14\frac{2}{3}$

Problem #8

8. $2\frac{1}{4}$ groups of $\frac{2}{5}$

Questions and Conversations for #8

See Questions and Conversations for #6.

Solution for #8

Value: $2\frac{1}{4}$ groups of $\frac{2}{5}$ equals $\frac{9}{10}$.

Sample diagram:

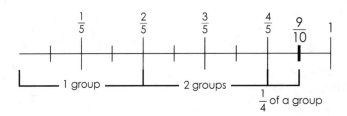

The number line shows that $\frac{1}{4}$ of a group of $\frac{2}{5}$ is $\frac{1}{10}$. So $2\frac{1}{4}$ groups of $\frac{2}{5}$ is $\frac{2}{5}+\frac{2}{5}+\frac{1}{10}=\frac{9}{10}$.

Sample story: A glass holds $\frac{2}{5}$ of a liter (0.4 L) of juice. How many liters will $2\frac{1}{4}$ glasses hold? (Answer: $\frac{9}{10}$ of a liter [0.9 L])

Number model: $\frac{2}{5} \cdot 2\frac{1}{4} = \frac{9}{10}$

Problem #9

9. Does the algorithm that you created in Problem #5 still work for Problems #6–#8? If so, show why. If not, try to modify it or create a new algorithm that will work. Explain your thinking.

Solution for #9

Students should find that the algorithm they discovered in Stage 1 still works for Problem #6.

$$5 \cdot \frac{3}{7} = \frac{5}{1} \cdot \frac{3}{7} = \frac{15}{7} = 2\frac{1}{7}$$

It will also work for Problems #7 and #8, provided that you rewrite the mixed numbers as fractions.

$$4 \cdot 3\frac{2}{3} = \frac{4}{1} \cdot \frac{11}{3} = \frac{44}{3} = 14\frac{2}{3}$$

$$\frac{2}{5} \cdot 2\frac{1}{4} = \frac{2}{5} \cdot \frac{9}{4} = \frac{18}{20} = \frac{9}{10}$$

Both of these agree with the results obtained from the drawings.

Some students find it more natural to decompose the mixed numbers into their two parts and multiply each part by the other number. (This is the *distributive property*.) They may still rely on the original algorithm to handle the individual parts.

$$4 \cdot 3\frac{2}{3} = 4 \cdot 3 + 4 \cdot \frac{2}{3} = 12 + 2\frac{2}{3} = 14\frac{2}{3}$$

$$\frac{2}{5} \cdot 2\frac{1}{4} = \frac{2}{5} \cdot 2 + \frac{2}{5} \cdot \frac{1}{4} = \frac{4}{5} + \frac{2}{20} = \frac{8}{10} + \frac{1}{10} = \frac{9}{10}$$

STAGE 3

Problem #10

10. $1\frac{3}{4}$ groups of $1\frac{1}{2}$

Solution for #10

Value: $1\frac{3}{4}$ groups of $1\frac{1}{2}$ equals $2\frac{5}{8}$.

Sample diagram:

In Step 1, we create a group of $1\frac{1}{2}$. In Step 2, we join another $\frac{3}{4}$ of this group to the original in order to make $1\frac{3}{4}$ groups of $1\frac{1}{2}$. In Step 3, we show 8 parts in each whole and label the values of the remaining two parts. The total value is

$$1 + \frac{1}{2} + \frac{3}{4} + \frac{3}{8} = 2\frac{5}{8}.$$

Sample story: A length of rope weighs $1\frac{1}{2}$ pounds per meter. How much will $1\frac{3}{4}$ meters of the rope weigh? (Answer: $2\frac{5}{8}$ pounds)

Number model: $1\frac{1}{2} \cdot 1\frac{3}{4} = 2\frac{5}{8}$

Teacher's Note. Our diagram is an area model, while the story suggests a number line model. Consider challenging students to create a number line drawing for this situation.

Problem #11

11. If you need to modify an old algorithm or create a new one for Problem #10, do this and explain your thinking. If not, show that one of your earlier algorithms works.

Solution for #11

The algorithm from Stage 1 will still work if you rewrite the mixed numbers as fractions.

$$1\frac{1}{2} \cdot 1\frac{3}{4} = \frac{3}{2} \cdot \frac{7}{4} = \frac{21}{8} = 2\frac{5}{8}$$

You can see this process in the diagram. The $2 \cdot 4$ in the denominator corresponds with finding the 8 equal parts in the each whole. The $3 \cdot 7$ in the numerator is connected to the 7 rows of 3 eighths.

Some students may use the equation $1 + \frac{1}{2} + \frac{3}{4} + \frac{3}{8} = 2\frac{5}{8}$ (above) to develop a different strategy based on the four parts in their drawing. The four numbers in the equation come from 1 group of 1, 1 group of $\frac{1}{2}$, $\frac{3}{4}$ of a group of 1, $\frac{3}{4}$ of a group of $\frac{1}{2}$ or $(1 \cdot 1) + (\frac{1}{2} \cdot 1) + (1 \cdot \frac{3}{4}) + (\frac{1}{2} \cdot \frac{3}{4})$.

This may remind students of the standard process for multiplying 2-digit numbers. Both are based on the *distributive property*.

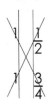

Making this connection may help students avoid the common error of multiplying only the two whole numbers and the two fractions before adding the results. This diagram shows that the whole number-fraction products would be missing.

WRAP UP

Share Strategies

Allow students to share and critique one another's diagrams and stories. Clarify important ideas and correct misconceptions. As students talk, ask them to think about:

» Which diagrams are clearest and easiest to understand? Why?

» Which diagrams and stories fit the expression the best? Why?

» Which stories are the most realistic or the most interesting? Why?

Summarize

Answer any remaining questions that students have. Summarize and expand on key ideas:

» $a \cdot b$ represents b groups of a. When b is less than 1, you might phrase things differently, but the essential meaning remains consistent. For example, although $6 \cdot 5$ means "5 groups of 6," you might read $6 \cdot \frac{2}{3}$ as "$\frac{2}{3}$ of a group of 6." You are changing the wording from "how many groups of 6" to "how much of a group of 6."

» When you multiply two numbers, you express the answer in terms of the size of the original whole.

» When you multiply mixed numbers, it is natural to decompose them and multiply the parts (the *distributive property*). You should know this strategy as well as the standard procedure of rewriting the mixed number(s) as fractions; it gives you the flexibility to choose the best approach for a given situation.

» Diagrams help you understand and explain why fraction multiplication procedures work.

Further Exploration

Ask students to think of new questions to ask or ways to extend this exploration. Here are some possibilities:

» Redo some of the Stage 2 problems, reversing the order of the numbers.

» Create diagrams and stories for larger mixed number expressions such as $3\frac{4}{5}$ groups of $2\frac{3}{8}$. Can you find a way to construct drawings effectively to illustrate both types of procedures (the distributive property and improper fractions)?

» Explore the following situations by choosing numbers, drawing diagrams, doing calculations, and creating stories. What patterns do you see? What causes them?

- b groups of $\dfrac{a}{b}$ (or $\dfrac{a}{b}$ groups of b)

- $\dfrac{a}{b}$ groups of $\dfrac{b}{c}$ (or $\dfrac{b}{c}$ groups of $\dfrac{a}{b}$)

Exploration  6
Undo It!

INTRODUCTION

Materials

- » Calculator (for Problem #1)
- » Graph paper (recommended for drawing diagrams in Stages 2 and 3)

Prior Knowledge

- » Understand fraction and decimal multiplication (meanings and procedures).
- » Complete Exploration 5: Visualizing Fraction Multiplication (recommended).
- » Have experience with "fact triangles" for multiplication and division. (Stages 2 and 3)

Learning Goals

- » Understand that reciprocals undo multiplication.
- » Understand reciprocals from multiple perspectives.
- » Create and flexibly use a variety of strategies for dividing fractions.
- » Connect the meanings of fraction multiplication and division.
- » Communicate complex mathematical ideas clearly.
- » Persist in solving challenging problems.

> **Teacher's Note.** This exploration works best before students have been exposed to the concept of a reciprocal and especially before they have learned a procedure for dividing fractions.

Launching the Exploration

Motivation and purpose. To students: The general concept of "undoing" is extremely important in mathematics. It shows up across multiple topics and at all levels, from the earliest grades to the most advanced college-level courses and beyond. In this activity, you will explore the idea of "undoing" multiplication, especially in the context of fractions.

Understanding the problem. Read the Motivation and Purpose section to students. Skim through the entire exploration, or as much of it as students will be completing, to get a sense for what is coming.

If students will be working on Stages 2 and 3, make sure they are familiar with multiplication/division fact triangles (see example below).

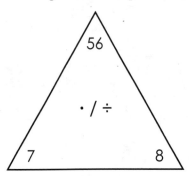

This triangle represents four connected equations, or "facts." In this case, $7 \cdot 8 = 56$, $8 \cdot 7 = 56$, $56 \div 7 = 8$, and $56 \div 8 = 7$.

To be consistent, the product (or the dividend) is always at the top of the triangle. Some students are surprised to find that these relationships continue to hold for numbers other than whole numbers.

Read Problem #1 with students to be sure they understand what they are being asked to do. Stress the importance of including both fractions and decimals in their responses.

Teacher's Note. See the Common Core State Standards for Mathematics (NGA & CCSSO, 2010) for a summary of different types of division situations. Although this exploration addresses a number of situations as they relate to fractions and the inverse relationship between multiplication and division, it does not attempt to cover every case. It is important to provide students with many opportunities to explore the various division situations over an extended period of time.

STUDENT HANDOUT

Stage 1

1. Find at least 10 pairs of numbers that make this equation true. Use a variety of numbers written in different forms. Describe your strategies.

$$6 \cdot \underline{} \cdot \underline{} = 6$$

What happens when you change the equation to $8 \cdot \underline{} \cdot \underline{} = 8$? Explain your thinking.

2. Make as many observations as you can about pairs of numbers that are solutions to equations of the form $a \cdot \underline{} \cdot \underline{} = a$. Explain the thinking behind your observations.

3. Find the missing factors in each equation. (Try not to use the same strategy each time.) Write a corresponding division equation for each.

   a. $\dfrac{3}{4} \cdot \dfrac{\square}{\square} = \dfrac{1}{2}$

   b. $\dfrac{2}{3} \cdot \dfrac{\square}{\square} = \dfrac{2}{5}$

   c. $\dfrac{\square}{\square} \cdot \dfrac{3}{4} = \dfrac{5}{6}$

For the triangles in Problems #4–#7,

» Find the missing value. Explain how you found it.

» Write the four equations that belong to the "fact family."

» Choose the division expression whose quotient belongs in place of the "?". Write a story for it.

» Draw a diagram that illustrates the division equation and your story.

Stage 2

4.

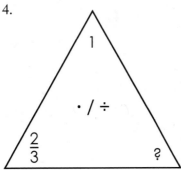

5.

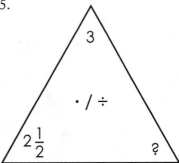

Stage 3

6.

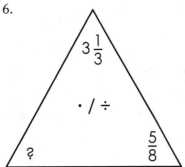

7.

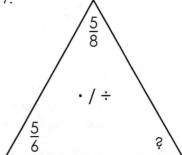

TEACHER'S GUIDE

STAGE 1

Problem #1

1. Find at least 10 pairs of numbers that make this equation true. Use a variety of numbers written in different forms. Describe your strategies.

$$6 \cdot \underline{\quad} \cdot \underline{\quad} = 6$$

What happens when you change the equation to $8 \cdot \underline{\quad} \cdot \underline{\quad} = 8$? Explain your thinking.

Questions and Conversations for #1

This section contains ideas for conversations, mainly in the form of questions that students may ask or that you may pose to them. Be sure to allow students to do most of the thinking and talking!

» *Is it okay to use a calculator?* Yes, especially when you are working with decimals. Decide when it is most helpful to use it and when it makes more sense to use mental math or paper and pencil methods.

» *Did the number in the first blank make the answer larger or smaller than 6? How can you adjust for this?* This is an important idea to pay attention to!

» *What happens if you try a fraction that does not have a numerator of 1?* This is more challenging than working with numbers like $\frac{1}{2}, \frac{1}{3}, \frac{1}{4}$, etc. Be sure to explore this!

» *What happens to the second number when you increase the number in the first blank?* Think about how you can compensate for the effect of the first number.

» *Are there any numbers that you can never put in the blanks?* Yes. Try to find one (or more).

» *Is it okay to try negative numbers?* Yes, if you know how to multiply them.

> **Teacher's Note.** Some students may have heard of reciprocals before and will discover that this is what they are looking for. Encourage them to write many of their answers in decimal form so that they can move beyond thinking of reciprocals as simply "flipping fractions." Ask them to look for a quick way to predict the value of the second decimal once they have chosen the first. (You can do this by dividing 1 by the first number).

93

Solution for #1

Sample responses:

$$6 \cdot 1 \cdot 1 = 6 \qquad\qquad 6 \cdot 2 \cdot 0.5 = 6$$

$$6 \cdot \frac{1}{3} \cdot 3 = 6 \qquad\qquad 6 \cdot 2.5 \cdot 0.4 = 6$$

$$6 \cdot 0.625 \cdot 1.6 = 6 \qquad\qquad 6 \cdot \frac{3}{4} \cdot 1.\overline{3} = 6$$

$$6 \cdot 1\frac{3}{4} \cdot \frac{4}{7} = 6 \qquad\qquad 6 \cdot \frac{3}{10} \cdot 3\frac{1}{3} = 6$$

$$6 \cdot -1 \cdot -1 = 6 \qquad\qquad 6 \cdot -\frac{4}{11} \cdot -2\frac{3}{4} = 6$$

Any pair of numbers whose product is 1 will work. Changing the equation to $8 \cdot \underline{} \cdot \underline{} = 8$ does not change the solutions!

In terms of strategies, students may observe that if their first number made the answer smaller than 6, then they had to find one that would increase it again, or vice versa. Some will estimate and use a trial and error process. (Calculators might be helpful in this case.) Some students may discover a more specific compensation strategy. For example, if they already know that 0.8 and 1.25 is a pair that works, they may realize they if they double 0.8, then they must compensate by taking half of 1.25. This gives the pair 1.6 and 0.625.

Students who discover that they can multiply their two chosen numbers before multiplying by 6 (using the *associative* property) may realize that they just need to find two numbers whose product is 1.

Problem #2

2. Make as many observations as you can about pairs of numbers that are solutions to equations of the form $a \cdot \underline{} \cdot \underline{} = a$. Explain the thinking behind your observations.

Teacher's Note. There are no responses shown for these questions, because they are intended just to summarize and focus students' thinking on issues that have (hopefully!) already come up in Problem #1.

Questions and Conversations for #2

» *What do the pairs have in common?*

» *Once you know one of the numbers in a pair, how can you find the other?*

» *Is it easier to recognize pairs when they are written in certain forms?*

» *Are there any numbers that can never be part of a pair?*

Solution for #2

Sample responses:

» The product of the numbers in every pair is equal to 1.

» For positive numbers, one of the numbers is always less than 1, and the other is always greater than 1 (unless they are both equal to 1).

» The smaller one number is, the larger the other is.

» The pairs are called *reciprocals* (regardless of whether they are written in fraction or decimal form).

» If you multiply *a* by some number *n*, then multiplying the result by the reciprocal of *n* undoes this by returning the value to *a*.

» It is easier to recognize reciprocals when they are written as fractions, because the numerators and denominators are interchanged.

» If you know one of the numbers, you can find the other by dividing 1 by it.

» The reciprocal of *n* can be expressed as $\frac{1}{n}$ for any nonzero value of *n*.

» Both numbers in a pair have the same sign (both positive or both negative).

» The number 0 cannot be in any pair (it has no *reciprocal*), because if you multiply *a* by 0, there is no way to undo this by multiplying again.

» These observations hold if *a* equals any number except 0. If $a = 0$, then the numbers in the blanks can be any (real) numbers whatsoever.

The product of the numbers should always be 1, because $(a \cdot b) \cdot c = a \cdot (b \cdot c)$ (the *associative* property of multiplication). The student's chosen numbers, *b* and *c*, must have a product of 1 so that the expression on the right will become $a \cdot 1$. Also, if one number is less than 1, the other must be greater than 1, because multiplying by a number between 0 and 1 makes the answer smaller, and to return it to the original value you must make it larger again.

Problem #3

3. Find the missing factors in each equation. (Try not to use the same strategy each time.) Write a corresponding division equation for each.

a. $\frac{3}{4} \cdot \dfrac{\square}{\square} = \frac{1}{2}$

b. $\frac{2}{3} \cdot \dfrac{\square}{\square} = \frac{2}{5}$

c. $\dfrac{\square}{\square} \cdot \frac{3}{4} = \frac{5}{6}$

Questions and Conversations for #3

» *Is the number on the right side of the equation greater or less than the known factor?* It is important to pay attention to the sizes of the fractions, not just the individual numerators and denominators. This gives you important information about the value of the missing factor.

» *Will it be easier to compare the two numbers if you rewrite one or both fractions in equivalent forms?* It might. This depends on the fractions themselves, how you choose to rewrite them, and your understanding of how fractions work.

»   *Can you change the values of both fractions in such a way that the same missing factor still satisfies the equation?* Yes. For example, some people find it helpful to change them both into whole numbers. Can you see a way to do this without affecting the value of the missing factor?

»   *Is it possible to apply some of the ideas from Problems #1 and #2?* Yes. For example, think about how you might be able to use the idea that the product of reciprocals equals 1.

Teacher's Note. We assume that students have not yet learned to "multiply by the reciprocal of the divisor," though some may guess this after completing Problems #1 and #2. Others will discover it as they work on Problem #3! Even when they do notice this, they may continue to choose other strategies. In the meantime, they are often very creative in developing methods that make sense to them!

Solution for #3

a. Original equation: $\frac{3}{4} \cdot \frac{2}{3} = \frac{1}{2}$;

Division equation: $\frac{1}{2} \div \frac{3}{4} = \frac{2}{3}$

Sample strategy: Find a fraction equivalent to $\frac{1}{2}$ whose numerator is a multiple of 3 and whose denominator is a multiple of 4. $\frac{6}{12}$ works well for this:

$$\frac{3}{4} \cdot \frac{\boxed{}}{\boxed{}} = \frac{6}{12}.$$

This makes it easy to see that the missing fraction is $\frac{2}{3}$.

b. Original equation: $\frac{2}{3} \cdot \frac{3}{5} = \frac{2}{5}$; Division equation: $\frac{2}{5} \div \frac{2}{3} = \frac{3}{5}$.

Sample strategy: Rewrite the fractions as equivalent fractions that have the same denominator:

$$\frac{10}{15} \cdot \frac{\boxed{}}{\boxed{}} = \frac{6}{15}.$$

This makes it easier to see that the answer is $\frac{6}{10}$ (or $\frac{3}{5}$) as large as the first number. Therefore, you need to multiply the first number by $\frac{3}{5}$.

c. Original equation: $\frac{10}{9} \cdot \frac{3}{4} = \frac{5}{6}$; Division equation: $\frac{5}{6} \div \frac{3}{4} = \frac{10}{9}$.

Sample strategy 1: Multiply both fractions by the same number so that they become whole numbers. Because $\frac{3}{4} \cdot 12 = 9$ and $\frac{5}{6} \cdot 12 = 10$, $\frac{\boxed{}}{\boxed{}} \cdot \frac{3}{4} = \frac{5}{6}$ becomes $\frac{\boxed{}}{\boxed{}} \cdot 9 = 10$.

The missing fraction is $\dfrac{10}{9}$.

Sample strategy 2: Temporarily make the $\dfrac{3}{4}$ equal to 1 by multiplying it by $\dfrac{4}{3}$. Then multiply this by $\dfrac{5}{6}$ to make the answer equal to $\dfrac{5}{6}$. Multiplying by $\dfrac{4}{3}$ and then $\dfrac{5}{6}$ is equivalent to multiplying by $\dfrac{5}{6} \cdot \dfrac{4}{3} = \dfrac{20}{18} = \dfrac{10}{9}$.

$$\frac{5}{6}\cdot\left(\frac{4}{3}\cdot\frac{3}{4}\right)=\left(\frac{5}{6}\cdot\frac{4}{3}\right)\cdot\frac{3}{4}=\frac{10}{9}\cdot\frac{3}{4}$$

Notice that any of these strategies will work for any of the questions—and that many other strategies are possible as well.

STAGE 2

For the triangles in Problems #4–#7 (Stages 2 and 3),
» Find the missing value. Explain how you found it.
» Write the four equations that belong to the "fact family."
» Choose the division expression whose quotient belongs in place of the "?". Write a story for it.
» Draw a diagram that illustrates the division equation and your story.

Problem #4

4.

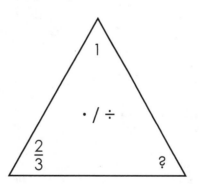

Questions and Conversations for #4

» *Is it important to do the four parts of each problem in a particular order?* Not necessarily. For example, if you think of a story first, it might guide you through the process of drawing the diagram. Or if you draw a picture first, this could help you find the missing value. In the end, your diagram should contain information that describes your story.

» *Might it help to create a multiplication story first and use it to develop a division story?* This is worth considering. Multiplication stories may be easier to create, and you can use the inverse relationship between multiplication and division to rewrite the story so that it fits division.

» *There are two important meanings of division based on the idea of grouping. What are they, and how can you apply this knowledge to help you create diagrams and stories?* (See Exploration 1: Sharing and Grouping for more information about this.) "How many (or much) in each group? " division involves splitting a number into groups of equal size and asking for the size of each group. "How many groups?" division is the reverse of this. You know the size of each group, and you want to know how many groups there are. As you create your stories or drawings, keep both of these possibilities in mind. In any given situation, one may be easier than the other. Ask yourself what the groups are, how many of them there are, and what their sizes are. Show this in a picture, or relate it to a real-world situation.

> **Teacher's Note.** "How many in each group?" division is often called *partitive* division, while "How many groups?" division is sometimes referred to as *quotative* or *measurement* division.

> **Teacher's Note.** This is an example of "How much in each group?" division because there are $1\frac{1}{2}$ cups of flour in each group (1 recipe). If students' stories involve the question "How many groups of $\frac{2}{3}$ are in 1?", they are using "How many groups?" division.

Solution for #4

The missing value is $\frac{3}{2}$. Typical strategies for finding this are similar to those shown in Problem #3. Some students may notice that because "1" is at the top of the triangle, they are looking for the reciprocal of $\frac{2}{3}$.

Fact family:

$$\frac{2}{3} \cdot \frac{3}{2} = 1 \qquad\qquad \frac{3}{2} \cdot \frac{2}{3} = 1$$

$$1 \div \frac{2}{3} = \frac{3}{2} \qquad\qquad 1 \div \frac{3}{2} = \frac{2}{3}$$

Sample story for $1 \div \frac{2}{3} = \frac{3}{2}$: You are using 1 cup of flour to make $\frac{2}{3}$ of a brownie recipe. How much flour did the original recipe call for?

Sample diagram:

Problem #5

5.

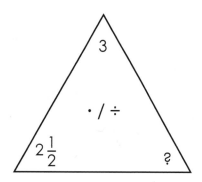

Questions and Conversations for #5

See Questions and Conversations for #4.

Solution for #5

The missing value is $1\frac{1}{5}$ or $\frac{6}{5}$. Typical strategies for finding this are similar to those shown in Problem #3. Comparing with Problem #4, students could also observe that if there were a "1" at the top of the triangle, the answer would be $\frac{2}{5}$ (the reciprocal of $2\frac{1}{2}$). Because there is a "3" at the top instead, the answer is 3 times greater than this, and $\frac{2}{5} \cdot 3 = \frac{6}{5}$. (This is another way to understand why the standard method of multiplying by the reciprocal works!)

Fact family:

$$2\frac{1}{2} \cdot 1\frac{1}{5} = 3 \qquad 1\frac{1}{5} \cdot 2\frac{1}{2} = 3$$

$$3 \div 2\frac{1}{2} = 1\frac{1}{5} \qquad 3 \div 1\frac{1}{5} = 2\frac{1}{2}$$

Sample story for $3 \div 2\frac{1}{2} = 1\frac{1}{5}$:
You can walk $2\frac{1}{2}$ miles in an hour. How long will it take you to walk 3 miles? (Answer: $1\frac{1}{5}$ hours, or 1 hour and 12 minutes)

Teacher's Note. This is an example of "How many groups?" division, because there are $1\frac{1}{5}$ groups of $2\frac{1}{2}$ in 3. If students' stories involve the question, "If there are 3 of something in $2\frac{1}{2}$ groups, then how much is in 1 group?", they are using "How many (much) in each group?" division.

Sample diagram:

	$\frac{1}{5}$	$\frac{2}{5}$	$\frac{3}{5}$	$\frac{4}{5}$	1	$1\frac{1}{5}$	Hours
0							
0	$\frac{1}{2}$	1	$1\frac{1}{2}$	2	$2\frac{1}{2}$	3	Miles walked

STAGE 3

Problem #6

6.

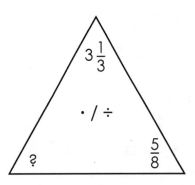

Questions and Conversations for #6

See Questions and Conversations for #4.

Solution for #6

The missing value is $5\frac{1}{3}$. Typical strategies for finding this are similar to those shown in Problem #3.

Fact family:

$$\frac{5}{8} \cdot 5\frac{1}{3} = 3\frac{1}{3} \qquad 5\frac{1}{3} \cdot \frac{5}{8} = 3\frac{1}{3}$$

$$3\frac{1}{3} \div \frac{5}{8} = 5\frac{1}{3} \qquad 3\frac{1}{3} \div 5\frac{1}{3} = \frac{5}{8}$$

Teacher's Note. This involves the "How many in each group?" meaning of division.

Sample story for $3\frac{1}{3} \div \frac{5}{8} = 5\frac{1}{3}$:

You have ridden your bike $3\frac{1}{3}$ miles, which is $\frac{5}{8}$ of the way from your house to your friend's house. How many miles does she live from you?

Sample diagram:

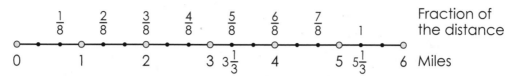

Problem #7

7.

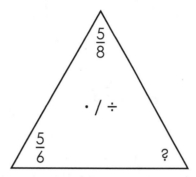

Questions and Conversations for #7

See Questions and Conversations for #4.

Solution for #7

The missing value is $\frac{3}{4}$. Typical strategies for finding this are similar to those shown in Problem #3.

Fact family:

$$\frac{5}{6} \cdot \frac{3}{4} = \frac{5}{8} \qquad \frac{3}{4} \cdot \frac{5}{6} = \frac{5}{8}$$

$$\frac{5}{8} \div \frac{5}{6} = \frac{3}{4} \qquad \frac{5}{8} \div \frac{3}{4} = \frac{5}{6}$$

Sample story for $\frac{5}{8} \div \frac{5}{6} = \frac{3}{4}$: You are serving ice cream to some friends at your birthday party. You have served $\frac{5}{6}$ of them so far and used $\frac{5}{8}$ of a gallon of the ice cream. Assuming that you serve the same amount to each person, how many gallons of ice cream will you use to serve them all?

> **Teacher's Note.** This involves the "How many in each group?" meaning of division.

Sample diagram:

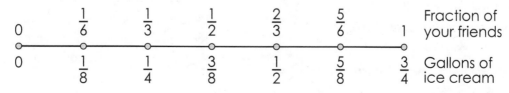

WRAP UP

Share Strategies

Allow students to share and critique one another's diagrams, stories, and strategies for finding missing factors. Clarify important ideas and correct any misconceptions. In Stages 2 and 3, determine whether students' stories represent the "How many groups?" or the "How many in each group?" meaning for division. (Most stories should fit into one of these two categories.)

Summarize

Answer any remaining questions that students have. Summarize, extend, and discuss a few key ideas:

» Reciprocals are about much more than "flipping fractions." They are numbers than can be used to undo the process of multiplication, and they can be written in fraction or decimal form. In mathematics, these types of numbers are called *inverses*. In fact, the word *inverse* applies to the general concept of undoing in mathematics. Reciprocals are often called *multiplicative inverses*.

» There are many strategies and procedures for dividing fractions. The standard method is to multiply the dividend by the reciprocal of the divisor. Does this procedure apply to decimals as well? (Yes, it must. Decimals and fractions are just different ways of representing numbers. The form in which you write a number should not change the result of a calculation. Encourage students to experiment with this.)

Further Exploration

Ask students to think of new questions to ask or ways to extend this exploration. Here are some possibilities:

» If you drew number lines for your diagrams in Problems #4–#7, look for the pairs of numbers that correspond with the number 1 in each drawing. What do you notice? (They are reciprocals. For example, in the solution to Problem #6, 1 mile corresponds to $\frac{3}{16}$ of the distance, while the (1) whole distance matches with $5\frac{1}{3}$, or $\frac{16}{3}$ miles. Why does this happen?)

» Think about matching your stories to other equations in the fact family. How would you have to rewrite them to make them fit?

» Create stories using different meanings for division. For example, if one of your stories was

Teacher's Note. This relationship should be apparent regardless of the type of diagram drawn. However, it might be easier to see with number lines.

based on "How many in each group?" division, try to write one using the idea of "How many groups?".

» Put other numbers in the fact triangles. Can you find combinations that produce new types of challenges?

Exploration  7

Sum-Product Pairs

INTRODUCTION

Prior Knowledge

» Know procedures for adding, subtracting, multiplying, and dividing positive and negative rational numbers.

» Have experience simplifying complex fractions (recommended).

» Understand the concept of equivalent algebraic expressions (Stage 3).

Learning Goals

» Use guess, check, and revise strategies to solve problems involving fractions.

» Increase computational fluency with the four basic operations on fractions, mixed numbers, negative numbers, and complex fractions.

» Identify and extend numerical patterns.

» Recognize and make use of symmetry in numerical contexts.

» Describe patterns and prove conjectures algebraically.

» Communicate complex mathematical ideas clearly.

» Persist in solving challenging problems.

Launching the Exploration

Motivation and purpose. To students: You may have noticed two cases in which the product of two numbers is equal to their sum: $0 + 0 = 0 \cdot 0$ and $2 + 2 = 2 \cdot 2$. Have you ever wondered if this happens with other numbers? There are some surprising and beautiful patterns connected with the idea. As you investigate them, you will strengthen your computational skills with fractions and negative numbers.

Understanding the problem. Discuss the introductory paragraph of the student handout to clarify the main ideas. Then look briefly through the entire activity. Students begin their work with whole numbers, then look for a pattern and apply it to extend the investigation to other types of numbers. In Stages 2 and 3, they explore deeper connections—to reciprocals and symmetry, in particular—and use algebra to show that the patterns they have observed apply in general.

In order to meet the learning goals of this exploration and to make the patterns visible, students need to write their numbers in fraction form. However, if they have a strong preference for decimals at first, you might allow students to use them for a while.

As they work, generate conversation about the advantages and disadvantages of each approach, gradually steering them toward fractions.

Students are often asked to show the details of their computational procedures in this exploration. Writing these out will help them develop fluency with adding, subtracting, multiplying, and dividing fractions and negative numbers. If they pay close attention to these processes as they work, it will also give them insight into the causes of the patterns they encounter.

NAME: _____ DATE: _____

STUDENT HANDOUT

In this investigation, you search for numbers that satisfy the equation $a + b = a \cdot b$. We will call these numbers "sum-product pairs" or just "partners" for short. The most familiar examples are $a = 0$, $b = 0$ and $a = 2$, $b = 2$. However, a and b need not be the same number. For example, $a = -1$, $b = \frac{1}{2}$ is a sum-product pair, because $-1 + \frac{1}{2} = -\frac{1}{2}$ and $-1 \cdot \frac{1}{2} = -\frac{1}{2}$. As you work, keep in mind the possibility that some numbers may not have partners!

Stage 1

1. Find a partner (b) for $a = 4$. Show your thinking process in detail, including some sample calculations.

2. Use what you learned in Problem #1 to help you look for partners for $a = 1$, $a = 3$, and $a = 5$. Again, describe your thinking process.

3. Organize your results for $a = 0, 1, 2, 3, 4,$ and 5. Find a pattern that makes it easier to predict b from a, and write it as a formula. Show the calculations to prove that your formula works for $a = 0, 1, 2, 3, 4,$ and 5.

4. Show how to use your pattern from Problem #3 to find a partner for each of the following 12 numbers. Show the details of at least one calculation from each set. Describe any new patterns that you notice.

Set 1: $\frac{2}{3}, \frac{3}{4}, \frac{4}{5}$ Set 2: $-\frac{2}{3}, -\frac{3}{4}, -\frac{4}{5}$

Set 3: $2\frac{1}{2}, 3\frac{1}{2}, 4\frac{1}{2}$ Set 4: $-2\frac{1}{2}, -3\frac{1}{2}, -4\frac{1}{2}$

5. Use your calculator to find an approximate value for the partner (b) of π. Show your process. Test your answer by calculating $\pi + b$ and $\pi \cdot b$. Calculate and write all of your numbers as precisely as your calculator allows.

Stage 2

6. Predict what will happen if you substitute a number's partner back into your formula. Explain your thinking. Test your prediction for the sum-product pair: 3, $\frac{3}{2}$. Then test it for one pair from Problem #4 and for π and its partner. Show your calculations.

7. Make a conjecture about what happens when you subtract 1 from both numbers in a sum-product pair. Test your conjecture on a variety of numbers, showing at least three examples. Does this suggest a new way to find sum-product pairs?

8. Consider the following reciprocal pairs:

   a. $\frac{2}{5}$ and $\frac{5}{2}$ b. 6 and $\frac{1}{6}$ c. -5 and $-\frac{1}{5}$ d. $-\frac{2}{7}$ and $-\frac{7}{2}$ e. -1 and -1

   Find the partner of each number in each pair. (You do not have to show your calculations.) Describe any patterns you see.

Stage 3

9. Multiply a by its partner. Then add a to its partner. Do you get the same algebraic expression for both? Show a numerical example to verify that this expression correctly predicts the sum and product of a partner pair.

10. Use algebraic expressions or equations to prove that your patterns and conjectures from Problems #6 and #7 hold true in general. (*Note.* It may be easier to do them in reverse order: #7 and #6.)

TEACHER'S GUIDE

In this investigation, you search for numbers that satisfy the equation $a + b = a \cdot b$. We will call these numbers "sum-product pairs" or just "partners" for short. The most familiar examples are $a = 0$, $b = 0$ and $a = 2$, $b = 2$. However, a and b need not be the same number. For example, $a = -1$, $b = \frac{1}{2}$ is a sum-product pair, because $-1 + \frac{1}{2} = -\frac{1}{2}$ and $-1 \cdot \frac{1}{2} = -\frac{1}{2}$. As you work, keep in mind the possibility that some numbers may not have partners!

STAGE 1

Problem #1

1. Find a partner (b) for $a = 4$. Show your thinking process in detail, including some sample calculations.

Questions and Conversations for #1

This section contains ideas for conversations, mainly in the form of questions that students may ask or that you may pose to them. Be sure to allow students to do most of the thinking and talking!

» *How can you organize your work?* One good option is to make a table with columns labeled a, b, $a + b$, and $a \cdot b$.

» *When you make a guess that doesn't work, how can you decide whether to try a larger or smaller number next?* Pay close attention to how the sum and the product compare. Which is larger? Are they getting closer together or farther apart?

> **Teacher's Note.** Remember that if students are using decimals, you will gradually want to guide them toward working with fractions and mixed numbers (except in Problem #5). Otherwise, they will not see the patterns that will help them move forward. They will appreciate the value of this when they begin to realize that the decimals of some partners will go on forever!

Solution for #1

$b = 1\frac{1}{3}$ or $\frac{4}{3}$. This table shows one way to organize the information. Most students will test more numbers than this before they find the correct partner.

a (chosen)	b (guess)	$a + b$	$a \cdot b$
4	3	7	12
4	2	6	8
4	1	5	4
4	$1\frac{1}{2}$ or $\frac{3}{2}$	$5\frac{1}{2}$ or $\frac{11}{2}$	6
4	$1\frac{1}{4}$ or $\frac{5}{4}$	$5\frac{1}{4}$ or $\frac{21}{4}$	5
4	$1\frac{1}{3}$ or $\frac{4}{3}$	$5\frac{1}{3}$ or $\frac{16}{3}$	$5\frac{1}{3}$ or $\frac{16}{3}$

Students will often test small whole numbers first. They might observe that 4's partner should be between 1 and 2 because $a + b$ is larger than $a \cdot b$ when $b = 1$ and is *smaller* than $a \cdot b$ when $b = 2$.

Problem #2

2. Use what you learned in Problem #1 to help you look for partners for $a = 1$, $a = 3$, and $a = 5$. Again, describe your thinking process.

Questions and Conversations for #2

»   *How can you apply what you learned in Problem #1?* You may be able to work faster, because you have some experience now. You might also see patterns that speed things up.

Solution for #2

Most students will probably continue to use "trial and error" strategies to discover that the partners for 3 and 5 are $\frac{3}{2}$ and $\frac{5}{4}$. However, they are likely to find these answers more easily because of the experience they gained in the first question. Some may already guess a pattern and use it to find the partners immediately.

The number 1 has no partner, because multiplying any number by 1 leaves it unchanged, while adding 1 makes it one greater.

Problem #3

3. Organize your results for $a = 0, 1, 2, 3, 4,$ and 5. Find a pattern that makes it easier to predict b from a, and write it as a formula. Show the calculations to prove that your formula works for $a = 0, 1, 2, 3, 4,$ and 5.

Questions and Conversations for #3

» *What do you do if a number doesn't have a partner?* You can just write "none" in the place where you are recording the partners.

» *Why might you not be seeing patterns?* Consider checking for errors. Or, again, if you are using decimals, try writing your answers as fractions. (If the numbers are greater than 1, it will make your work easier if you write them as improper fractions.)

» *How can you say that the formula "works" for a number that doesn't have a partner?* The formula will show you this by not giving you an answer!

Solution for #3

Many students will probably use a table to organize their results:

a	b	Sum/Product
0	0	0
1	none	none
2	2	4
3	$\frac{3}{2}$	$4\frac{1}{2}$ or $\frac{9}{2}$
4	$\frac{4}{3}$	$5\frac{1}{3}$ or $\frac{16}{3}$
5	$\frac{5}{4}$	$6\frac{1}{4}$ or $\frac{25}{4}$

There are two common ways to show the pattern with a formula: $\frac{a}{a-1}$ or $1+\frac{1}{a-1}$.

The first expression is easier for most students.

0: $\dfrac{0}{0-1} = \dfrac{0}{-1} = 0$

1: $\dfrac{1}{1-1}$ would be $\dfrac{1}{0}$, but this is undefined, so it gives no answer.

2: $\dfrac{2}{2-1} = \dfrac{2}{1} = 2$

3: $\dfrac{3}{3-1} = \dfrac{3}{2}$

4: $\dfrac{4}{4-1} = \dfrac{4}{3}$

5: $\dfrac{5}{5-1} = \dfrac{5}{4}$

Problem #4

4. Show how to use your pattern from Problem #3 to find a partner for each of the following 12 numbers. Show the details of at least one calculation from each set. Describe any new patterns that you notice.

$$\text{Set 1: } \frac{2}{3}, \frac{3}{4}, \frac{4}{5} \qquad \text{Set 2: } -\frac{2}{3}, -\frac{3}{4}, -\frac{4}{5}$$

$$\text{Set 3: } 2\frac{1}{2}, 3\frac{1}{2}, 4\frac{1}{2} \qquad \text{Set 4: } -2\frac{1}{2}, -3\frac{1}{2}, -4\frac{1}{2}$$

Questions and Conversations for #4

» *Why are the numbers separated into sets?* Each set shows a distinctive pattern.

» *What makes these partners more difficult to find than the ones for whole numbers?* Some involve working with negative numbers. Others involve complex fractions.

» *How does having a negative number in the numerator and/or denominator affect the value of a fraction?* Think of the fraction as a division expression and apply your knowledge of division with negative numbers.

» *How do you handle complex fractions when they arise?* Apply what you know about connections between fractions and division or about processes that produce equivalent fractions, in order to write them in a simpler form. (See Exploration 1.)

» *Can you analyze your calculations to see what causes the patterns?* Yes, the regularity in the calculation processes might help you see this.

» *Are there other sets of numbers whose partners display similar patterns?* Yes. Think about the features that the fractions in the given sets have in common. How could you change the details or create sets with different common features? (For example, students might consider fractions whose numerators and denominators differ by 2 instead of 1, or mixed numbers whose fractional part is $\frac{1}{3}$ instead of $\frac{1}{2}$.)

» *How can you check that you have calculated each partner correctly?* Find its product and its sum. Verify that the two are equal.

Teacher's Note. If students understand *opposites* and *reciprocals* as numbers that undo addition and multiplication respectively, consider using the language *additive inverse* (for *opposite*) and *multiplicative inverse* (for *reciprocal*).

Solution for #4

Set 1: The partners of $\frac{2}{3}$, $\frac{3}{4}$, and $\frac{4}{5}$ are -2, -3, and -4 respectively.

Sample calculation for $\frac{4}{5}$:

$$\frac{\frac{4}{5}}{\frac{4}{5}-1} = \frac{\frac{4}{5}}{-\frac{1}{5}} = \frac{4}{5} \div -\frac{1}{5} = \frac{4}{5} \cdot -\frac{5}{1} = -4$$

Each partner is just the opposite of the numerator!

Set 2: The partners of $-\dfrac{2}{3}$, $-\dfrac{3}{4}$, and $-\dfrac{4}{5}$ are $\dfrac{2}{5}$, $\dfrac{3}{7}$, and $\dfrac{4}{9}$ respectively.

Sample calculation for $-\dfrac{4}{5}$: $\dfrac{-\dfrac{4}{5}}{-\dfrac{4}{5}-1} = \dfrac{-\dfrac{4}{5}}{-\dfrac{9}{5}} = -\dfrac{4}{5} \div -\dfrac{9}{5} = -\dfrac{4}{5} \cdot -\dfrac{5}{9} = \dfrac{4}{9}$

The partner is the positive fraction whose numerator is the same as the numerator of the original fraction and whose denominator is the sum of the numerator and denominator of the original fraction.

Set 3: The partners of $2\dfrac{1}{2}$, $3\dfrac{1}{2}$, and $4\dfrac{1}{2}$ are $1\dfrac{2}{3}$, $1\dfrac{2}{5}$, and $1\dfrac{2}{7}$ respectively.

Sample calculation for $4\dfrac{1}{2}$: $\dfrac{4\dfrac{1}{2}}{4\dfrac{1}{2}-1} = \dfrac{4\dfrac{1}{2}}{3\dfrac{1}{2}} = \dfrac{9}{2} \div \dfrac{7}{2} = \dfrac{9}{2} \cdot \dfrac{2}{7} = \dfrac{9}{7} = 1\dfrac{2}{7}$

Each partner's whole number is 1, and its numerator is 2. Its denominator is 1 less than twice the whole number part of the original mixed number.

Set 4: The partners of $-2\dfrac{1}{2}$, $-3\dfrac{1}{2}$, and $-4\dfrac{1}{2}$ are $\dfrac{5}{7}$, $\dfrac{7}{9}$, and $\dfrac{9}{11}$ respectively.

Sample calculation for $-4\dfrac{1}{2}$: $\dfrac{-4\dfrac{1}{2}}{-4\dfrac{1}{2}-1} = \dfrac{-4\dfrac{1}{2}}{-5\dfrac{1}{2}} = -\dfrac{9}{2} \div -\dfrac{11}{2} = -\dfrac{9}{2} \cdot -\dfrac{2}{11} = \dfrac{9}{11}$

Each partner's numerator is 1 more than twice the opposite of the integer part of the mixed number. The partner's denominator is 2 greater than its numerator.

Problem #5

5. Use your calculator to find an approximate value for the partner (b) of π. Show your process. Test your answer by calculating $\pi + b$ and $\pi \cdot b$. Calculate and write all of your numbers as precisely as your calculator allows.

> **Teacher's Note.** Draw students' attention to the fact that a single pattern (the one discovered in Problem #3) appears as many different patterns "in disguise!" Also, encourage them to check their results from time to time by finding sums and products of the partners to verify that they are equal. For example, to verify that $-\dfrac{4}{5}$ and $\dfrac{4}{9}$ are a sum-product pair:
>
> $-\dfrac{4}{5} + \dfrac{4}{9} = -\dfrac{36}{45} + \dfrac{20}{45} = -\dfrac{16}{45}$ and $-\dfrac{4}{5} \cdot \dfrac{4}{9} = -\dfrac{16}{45}$.

113

Solution for #5

$$b = \frac{a}{a-1} = \frac{\pi}{\pi-1} \approx \frac{3.14159265359}{3.14159265359-1} = \frac{3.14159265359}{2.14159265359} \approx 1.46694220692$$

$$\pi + b \approx 3.14159265359 + 1.46694220692 \approx 4.60853486051$$

$$\pi \cdot b \approx 3.14159265359 \cdot 1.46694220692 \approx 4.60853486051$$

STAGE 2

Problem #6

6. Predict what will happen if you substitute a number's partner back into your formula. Explain your thinking. Test your prediction for the sum-product pair: 3, $\frac{3}{2}$. Then test it for one pair from Problem #4 and for π and its partner. Show your calculations.

Questions and Conversations for #6

» *Can number patterns show symmetry? What might this look like?* Yes. For example, place values show symmetry about the ones place. The commutative property describes a symmetrical relationship between numbers.

» *How might the two numbers in a sum-product pair exhibit symmetry?* Consider opposites and reciprocals as examples. For instance, if you take the opposite of 4, and then take the opposite of that, you get the number 4 back again. Thus, -4 and 4 are opposites of each other. The same thing happens with reciprocals. Think about how this applies to sum-product pairs.

Solution for #6

You should get the original number back again, because they are partners of each other! For example, if you substitute 3's partner ($\frac{3}{2}$) into the formula, you should get an answer of 3.

$$\frac{\frac{3}{2}}{\frac{3}{2}-1} = \frac{\frac{3}{2}}{\frac{1}{2}} = \frac{3}{2} \div \frac{1}{2} = 3$$

It works!

Sample test for a pair from Problem #4: Earlier, we calculated that $\frac{4}{9}$ is the partner of $-\frac{4}{5}$. We can also show that $-\frac{4}{5}$ is the partner of $\frac{4}{9}$!

$$\frac{\frac{4}{9}}{\frac{4}{9}-1} = \frac{\frac{4}{9}}{-\frac{5}{9}} = \frac{4}{9} \div -\frac{5}{9} = \frac{4}{9} \cdot -\frac{9}{5} = -\frac{4}{5}$$

A test for π:

$$\frac{1.46694220692}{1.46694220692-1} = \frac{1.46694220692}{0.46694220692} \approx 3.14159265359$$

Problem #7

7. Make a conjecture about what happens when you subtract 1 from both numbers in a sum-product pair. Test your conjecture on a variety of numbers, showing at least three examples. Does this suggest a new way to find sum-product pairs?

Questions and Conversations for #7

» *Is a conjecture a random guess? What does this tell you about how you should make your conjecture?* No, a conjecture is not a random guess. It should be a prediction based on experience or knowledge—an "educated guess." Before making your conjecture, you should experiment with some numbers and make observations.

Solution for #7

Some examples:

Sum-Product Pair	Subtract 1 From Each
$4, \frac{4}{3}$	$3, \frac{1}{3}$
$2, 2$	$1, 1$
$-\frac{3}{4}, \frac{3}{7}$	$-\frac{7}{4}, -\frac{4}{7}$
$3\frac{1}{2}, 1\frac{2}{5}$	$\frac{5}{2}, \frac{2}{5}$

Based on these results, you may conjecture that when you subtract 1 from both numbers in a sum-product pair, you will get reciprocals. This suggests that an easy way to generate a sum-product pair is to create a reciprocal pair and just add 1 to each number. For example, if you begin with the reciprocal pair 2, $\frac{1}{2}$ and add one to each number, you get the sum-product pair 3, $\frac{3}{2}$.

Problem #8

8. Consider the following reciprocal pairs:

   a. $\frac{2}{5}$ and $\frac{5}{2}$ b. 6 and $\frac{1}{6}$ c. -5 and $-\frac{1}{5}$ d. $-\frac{2}{7}$ and $-\frac{7}{2}$ e. -1 and -1

   Find the partner of each number in each pair. (You do not have to show your calculations.) Describe any patterns you see.

Teacher's Note. In my experience, students often do not think to look at the sums of the fractions. This may be because they may have a restricted idea of what constitutes a pattern. Also, they tend to focus on the numerators and denominators separately. Encourage them to think of each fraction as a single number.

Questions and Conversations for #8

» *What should you do if many, but not all, of the pairs seem to show a pattern? Consider the possibility that you have made an error in your calculations.*

Solution for #8

The sum-product pairs for the reciprocals are:

a. $-\frac{2}{3}, \frac{5}{3}$ b. $\frac{6}{5}, -\frac{1}{5}$ c. $\frac{5}{6}, \frac{1}{6}$ d. $\frac{2}{9}, \frac{7}{9}$ e. $\frac{1}{2}, \frac{1}{2}$

Each pair has a sum of 1! (Students may see other patterns as well.)

STAGE 3

Problem #9

9. Multiply a by its partner. Then add a to its partner. Do you get the same algebraic expression for both? Show a numerical example to verify that this expression correctly predicts the sum and product of a partner pair.

Solution for #9

$$a \cdot \frac{a}{a-1} = \frac{a}{1} \cdot \frac{a}{a-1} = \frac{a^2}{a-1}$$

$$a+\frac{a}{a-1}=\frac{a\cdot(a-1)}{a-1}+\frac{a}{a-1}=\frac{a^2-a}{a-1}+\frac{a}{a-1}=\frac{a^2-a+a}{a-1}=\frac{a^2}{a-1}$$

Yes, the expressions agree. The result, $\frac{a^2}{a-1}$, is also consistent with what we found in Problem #3. For example, we found the sum and product for the pair 5, $\frac{5}{4}$ to be $\frac{25}{4}$. For $a=5$, this is $\frac{a^2}{a-1}=\frac{5^2}{5-1}=\frac{25}{4}$.

Problem #10

10. Use algebraic expressions or equations to prove that your patterns and conjectures from Problems #6 and #7 hold true in general. (*Note.* It may be easier to do them in reverse order: #7 and #6.)

Solution for #10

Sample response for a proof of Problem #6:

$$\frac{\frac{a}{a-1}}{\frac{a}{a-1}-1}=\frac{\frac{a}{a-1}}{\frac{a}{a-1}-1}\cdot\frac{a-1}{a-1}=\frac{a}{a-(a-1)}=\frac{a}{1}=a$$

Notice that each appearance of a in the original expression $\frac{a}{a-1}$ was replaced by the expression itself. When you simplify this, the result is a. This shows that the partner of a's partner is a—in other words, the relationship is symmetrical!

> **Teacher's Note.** Students who have not been taught procedures for working with algebraic fractions may still benefit from exploring Stage 3. Consider beginning with the proofs of Problems #8 and #7, because they are less complex. Structure them as group tasks and/or class discussions in order to provide students with the necessary support. They will learn a great deal from applying their knowledge of numeric fractions (creating equivalent fractions, finding common denominators, etc.) in an algebraic context. Resist the urge to teach rules. Instead, encourage students to develop, compare, test, and justify strategies with other students. You are likely to be impressed by the progress they make!

> **Teacher's Note.** If students are not sure how to handle expressions such as $a\cdot(a-1)$ because they have not yet seen the distributive property in an algebraic context, consider connecting it to familiar multiplication strategies that make use of it. For example, $50\cdot49=50\cdot(50-1)=50\cdot50-50\cdot1$.

117

Sample response for a proof of Problem #7:

1 less than a: $a - 1$

1 less than $\dfrac{a}{a-1}$: $\dfrac{a}{a-1} - 1 = \dfrac{a}{a-1} - \dfrac{a-1}{a-1} = \dfrac{a-(a-1)}{a-1} = \dfrac{1}{a-1}$

The two results are reciprocals, as expected.

Teacher's Note. Students do not need to apply the distributive property to understand the expression $a-(a-1)$. Instead, they can use their common sense! When the number they subtract is 1 less than the number they start with, the answer is always 1.

WRAP UP

Share Strategies

Give students an opportunity to discuss their methods for finding the original sum-product pairs. Ask them to describe and compare the patterns that they found.

Summarize

Answer any remaining questions that students have. Summarize and expand on a few key ideas:

» You can use algebraic expressions to represent numeric patterns. Equivalent algebraic expressions can represent the same pattern.

» You can use algebraic processes to prove conjectures about numbers.

» The symmetric relationship between the numbers in a sum-product pair reflects the symmetry in the commutative property of addition and multiplication: $a + b = b + a$ and $a \cdot b = b \cdot a$. The expressions remain the same when you reverse the roles of a and b.

Further Exploration

Ask students to think of new questions to ask or ways to extend this exploration. Here are some possibilities:

» If you know how to solve simple algebraic equations, substitute $a = 4$ into the equation $a + b = a \cdot b$ and then solve it for b. Repeat this process for different values of a; include some negative numbers, fractions, and mixed numbers. After you have experimented with this for a while, use what you have learned to solve the equation for b without substituting a number for a.

» See Problem #8. What happens when you look at reciprocals of partners instead of partners of reciprocals? (Answer: Both show the same pattern! The sum is always 1).

» Can you define pairs using other operations? For example, does it make sense to talk about "difference-quotient pairs"? Explain.

» When you add 1 to reciprocals, you get sum-product pairs. What do you get if you subtract 1 from reciprocals? What if you add or subtract 2, 3, 4, etc.?

» Can you find other algebraic expressions (besides $\frac{a}{a-1}$, $\frac{1}{a}$, and $-a$) that can be used to define symmetric pairs?

» Opposites are inverses for addition, because adding the opposite of a number "undoes" adding that number. Similarly, reciprocals are inverses for multiplication. Are sum-product pairs the inverses for some operation?

Teacher's Note. The answer is yes! Let the symbol ($\lozenge$) stand for the operation $x \lozenge y = x + y - x \cdot y$—call it "x *plimes* y," because it is a mixture of "plus" and "times." Most students probably will not discover this operation themselves, but if you give it to them, they can explore it. Challenge them to show that they can undo x "plimes" any number (except 1!) by "pliming" the result by its sum-product partner! What happens with $x \lozenge 1$?

Exploration 8

Unit Fraction Hunt

Prior Knowledge

- » Know procedures for adding, subtracting, multiplying, and dividing whole numbers, fractions, and decimals.
- » Know the meanings of exponents and square roots.
- » Understand and work with complex fractions.

Learning Goals

- » Use computational patterns to develop a deeper understanding of number relationships (especially with regard to fractions).
- » Increase general computational fluency with rational numbers.
- » Translate flexibly between different representations for numbers.
- » Apply properties of both numbers and operations to solve problems.
- » Use algebraic notation to represent and analyze number patterns.
- » Persist in solving challenging problems.

Launching the Exploration

Motivation and purpose. To students: This is an open-ended exploration that takes time and determination. Many students work on it for the better part of a school year! Even if you never finish it, it is fun to see how far you can get. If you work hard at it, you will become more fluent and flexible with numbers (especially fractions) and will improve your understanding of operations and number patterns—no matter how much of it you complete!

Understanding the problem. Read the Motivation and Purpose selection above to students. Then begin discussing the details of the exploration as outlined below.

Discuss the definition of a *unit fraction*: a fraction whose numerator is 1 and whose denominator is a nonzero whole number. The main goal of this exploration is to complete as much of the Unit Fraction Grid (see the last page of the handout) as possible by creating each unit fraction at the top of the grid using exactly four copies of the number on the left.

Read through the first page of the handout with students to help them understand the details. Explain that the number in the upper right-hand corner of each cell indi-

121

cates its approximate difficulty level. Tell students: It will help you make decisions as you work, but shouldn't be taken too seriously. You might occasionally spend a long time on a Stage 1 cell, or find the answer to a Stage 3 cell quickly. It depends on your mathematical background, your thinking style, and a little bit of luck!

Check students' understanding of the problem by having them begin to search for solutions to a few cells. Keep an eye on the Questions and Conversations listed in the Teacher's Guide. Some of the questions in the guide may surface right as students get started.

Teacher's Note. Consider using this exploration differently than others. I assign it as an optional challenge as soon as students have the necessary knowledge of rational number computation (including negative numbers)—leaving as much time as possible for interested students to work on it. In my experience, some work for a week or two and finish most of Stage 1 and some of Stage 2. For a few students, it becomes a mission! They continue working on it for the remainder of the school year. Typically, I have a student every 2 or 3 years who is able to complete the entire grid, and quite a few who come close!

STUDENT HANDOUT

Use exactly four copies of the number in the left column of the Unit Fraction Grid to make each unit fraction in the top row. You may not use any other numbers.

Example: In the "1" row and "$\frac{1}{3}$" column, you must use exactly four 1s (and no other numbers) to create an expression having a value of $\frac{1}{3}$. One possibility is $\frac{1}{1+1+1}$.

You may use:
 » addition, subtraction, multiplication, and division;
 » fraction bars and decimal points (including the bar to show repeating digits);
 » exponents and square roots;
 » negative numbers, parentheses, and the percent symbol; and
 » numbers as digits in larger numerals (such as two fives to make 55).

  Note. To get the most out of this exploration, challenge yourself to solve it without a calculator! This will force you to develop strategies that build deeper understandings of fractions and decimals, especially when you reach the more challenging cells. Stages 1 through 3 are marked in the upper right corner of each cell. The numbers indicate the level of difficulty of the cell based on the author's experience using the activity with students. Each individual's experience will be different.

Stage 1

Cells labeled with a "1" are often solved earlier in the process.

Stage 2

Cells labeled with a "2" may take more time and effort.

Stage 3

Cells labeled with a "3" are quite challenging for most people. Some of them are rarely found.

Additional Challenge

 » Find a pattern for each column. Use algebra to show why the patterns work.

UNIT FRACTION GRID

	$\frac{1}{2}$	$\frac{1}{3}$	$\frac{1}{4}$	$\frac{1}{5}$	$\frac{1}{6}$	$\frac{1}{7}$	$\frac{1}{8}$	$\frac{1}{9}$
1	1	1	3	1	3	3	2	2
2	2	1	1	2	2	3	1	2
3	1	1	1	2	1	3	1	1
4	1	1	2	1	2	2	1	2
5	2	1	1	2	1	3	2	2
6	2	1	3	1	2	1	2	2
7	2	1	3	2	1	1	1	2
8	2	1	1	2	3	1	2	1
9	2	1	2	2	2	2	1	2

TEACHER'S GUIDE

Use exactly four copies of the number in the left column of the Unit Fraction Grid to make each unit fraction in the top row. You may not use any other numbers.

Example: In the "1" row and "$\frac{1}{3}$" column, you must use exactly four 1s (and no other numbers) to create an expression having a value of $\frac{1}{3}$. One possibility is $\frac{1}{1+1+1}$.

You may use:

» addition, subtraction, multiplication, and division;

» fraction bars and decimal points (including the bar to show repeating digits);

» exponents and square roots;

» negative numbers, parentheses, and the percent symbol; and

» numbers as digits in larger numerals (such as two fives to make 55).

Note. To get the most out of this exploration, challenge yourself to solve it without a calculator! This will force you to develop strategies that build deeper understandings of fractions and decimals, especially when you reach the more challenging cells. Stages 1 through 3 are marked in the upper right corner of each cell. The numbers indicate the level of difficulty of the cell based on the author's experience using the activity with students. Each individual's experience will be different.

STAGE 1

Problems for Stage 1

See grid below. Cells labeled with a "1" are often solved earlier in the process.

	$\frac{1}{2}$	$\frac{1}{3}$	$\frac{1}{4}$	$\frac{1}{5}$	$\frac{1}{6}$	$\frac{1}{7}$	$\frac{1}{8}$	$\frac{1}{9}$
1	1	1	3	1	3	3	2	2
2	2	1	1	2	2	3	1	2
3	1	1	1	2	1	3	1	1
4	1	1	2	1	2	2	1	2
5	2	1	1	2	1	3	2	2

6	2	1	3	1	2	1	2	2
7	2	1	3	2	1	1	1	2
8	2	1	1	2	3	1	2	1
9	2	1	2	2	2	2	1	2

Questions and Conversations for Stage 1

This section contains ideas for conversations, mainly in the form of questions that students may ask or that you may pose to them. Be sure to allow students to do most of the thinking and talking!

The questions and notes in each stage of this exploration are typical of the ideas that you or the students might discuss at that point in their work. However, most of these questions could arise at any time.

» *Does every cell have a solution?* Yes. In fact, most, if not all, cells have more than one solution.

» *Can you use an equivalent fraction to solve a cell (for example, $\frac{2}{6}$ to solve a "$\frac{1}{3}$" cell)?* Absolutely! Equivalent fractions represent the same number. In fact, this is the only way to solve many of the cells.

» *When you solve one cell, can it ever help you solve another one?* Yes. There are many patterns on the grid. Consider looking at columns, diagonals, and even "L"-shaped patterns (like a knight's move in chess).

» *Can you square a number by writing it with an exponent of 2?* No—unless you are filling a cell in the "2" row. In this case, the exponent will count as one of your four 2s.

Teacher's Note. See Stage 3 for samples of additional solutions that do not appear in the grid.

Solution for Stage 1

See the next page of this section for sample solutions to the entire Unit Fraction Grid. Because each cell typically has many solutions, students will have answers that differ from the ones shown.

As you examine the solutions, watch for entries that could serve as the basis for a pattern in a column or diagonal. Look closely! Some of these patterns may be fairly easy to see in the grid, while others may only be hinted at by an entry in a single cell.

	$\dfrac{1}{2}$	$\dfrac{1}{3}$	$\dfrac{1}{4}$	$\dfrac{1}{5}$	$\dfrac{1}{6}$	$\dfrac{1}{7}$	$\dfrac{1}{8}$	$\dfrac{1}{9}$
1	$\dfrac{1\cdot 1}{1+1}$	$\dfrac{1}{1+1+1}$	$(1+1)^{-(1+1)}$	$\dfrac{1+1}{1\div .1}$	$\dfrac{.\overline{1}\cdot 1}{1-\sqrt{.\overline{1}}}$	$\dfrac{.\overline{1}}{1-(.\overline{1}+.\overline{1})}$	$\dfrac{.1}{1-(.1+.1)}$	$\dfrac{.\overline{1}}{1}\cdot\dfrac{1}{1}$
2	$\dfrac{2}{\sqrt{2}\cdot\sqrt{2}\cdot 2}$	$\dfrac{2}{2+2+2}$	$\dfrac{2}{2^2\cdot 2}$	$\dfrac{.2}{2^{2-2}}$	$\dfrac{.2}{2/2+.2}$	$\dfrac{.\overline{2}}{2-(.\overline{2}+.\overline{2})}$	$\dfrac{2}{2^{2\cdot 2}}$	$\dfrac{\frac{2}{2}}{2}\cdot .\overline{2}$
3	$\dfrac{3}{3\cdot 3-3}$	$\dfrac{3}{3+3+3}$	$\dfrac{3}{3\cdot 3+3}$	$\dfrac{.\overline{3}+.\overline{3}}{3+.\overline{3}}$	$\dfrac{3/3}{3+3}$	$\dfrac{.\overline{3}}{3-(.\overline{3}+.\overline{3})}$	$\dfrac{3}{3^3-3}$	$\dfrac{3}{3\cdot 3\cdot 3}$
4	$\dfrac{4+4}{4\cdot 4}$	$\dfrac{4}{4+4+4}$	$\dfrac{4\div 4}{\sqrt{4\cdot 4}}$	$\dfrac{4}{4\cdot 4+4}$	$\dfrac{\sqrt{4}}{4\cdot 4-4}$	$\dfrac{\sqrt{4}}{4\cdot 4-\sqrt{4}}$	$\dfrac{4}{(4+4)\cdot 4}$	$\dfrac{\sqrt{4}}{4\cdot 4+\sqrt{4}}$
5	$\dfrac{\sqrt{5}\cdot\sqrt{5}}{5+5}$	$\dfrac{5}{5+5+5}$	$\dfrac{5}{5\cdot 5-5}$	$\dfrac{5^{5-5}}{5}$	$\dfrac{5}{5\cdot 5+5}$	$\dfrac{.\overline{5}}{5-(.\overline{5}+.\overline{5})}$	$\dfrac{.5}{5-(.5+.5)}$	$\dfrac{.\overline{5}}{5}\cdot\dfrac{5}{5}$
6	$\dfrac{\sqrt{6}\cdot\sqrt{6}}{6+6}$	$\dfrac{6}{6+6+6}$	$\dfrac{6/6}{.\overline{6}\cdot 6}$	$\dfrac{6}{6\cdot 6-6}$	$\dfrac{6/6}{\sqrt{6}\cdot\sqrt{6}}$	$\dfrac{6}{6\cdot 6+6}$	$\dfrac{.6}{6-(.6+.6)}$	$\dfrac{.\overline{6}}{6}\cdot\dfrac{6}{6}$
7	$\dfrac{\sqrt{7}\cdot\sqrt{7}}{7+7}$	$\dfrac{7}{7+7+7}$	$\dfrac{.\overline{7}}{\sqrt{.\overline{7}}\cdot\sqrt{7}+.\overline{7}}$	$\dfrac{.7}{7}+\dfrac{.7}{7}$	$\dfrac{7}{7\cdot 7-7}$	$\dfrac{7^{7-7}}{7}$	$\dfrac{7}{7\cdot 7+7}$	$\dfrac{.\overline{7}}{7}\cdot\dfrac{7}{7}$
8	$\dfrac{\sqrt{8}\cdot\sqrt{8}}{8+8}$	$\dfrac{8}{8+8+8}$	$\dfrac{8+8}{8\cdot 8}$	$\dfrac{\sqrt{8\cdot 8}}{8}-.8$	$\dfrac{\sqrt{8\cdot .\overline{8}}}{8+8}$	$\dfrac{8}{8\cdot 8-8}$	$\dfrac{8/8}{\sqrt{8\cdot 8}}$	$\dfrac{8}{8\cdot 8+8}$
9	$\dfrac{\sqrt{9}\cdot\sqrt{9}}{9+9}$	$\dfrac{9}{9+9+9}$	$\dfrac{\sqrt{9}}{9\div\sqrt{9}+9}$	$\dfrac{\sqrt{9}}{9+9-\sqrt{9}}$	$\dfrac{9}{(9+9)\cdot\sqrt{9}}$	$\dfrac{\sqrt{9}}{\sqrt{9}+9+9}$	$\dfrac{9}{9\cdot 9-9}$	$\dfrac{9^{9-9}}{9}$

STAGE 2

Problems for Stage 2

See grid under Stage 1. Cells labeled with a "2" may take more time and effort.

> **Teacher's Note.** One way to do this is use the square root. For example, you can write 5 as either $\sqrt{5}\cdot\sqrt{5}$ or $\sqrt{5\cdot 5}$.

Questions and Conversations for Stage 2

» *What can you do if you are able to solve the cell with three copies of the number instead of four?* Think of a way to use two copies of a number to make the same number!

Teacher's Note. A factorial is written with an exclamation point. It is used for counting combinations and permutations of objects. You calculate it by multiplying a number by every natural number less than or equal to it. For example, $4! = 4 \cdot 3 \cdot 2 \cdot 1 = 24$. (The "1" isn't actually needed, of course.)

I do not usually mention the factorial until students bring it up (someone usually does!), but if they have been working for quite a while, the extra option may give them a new burst of energy.

Teacher's Note. If students want to use negative exponents but are unfamiliar with them, you might introduce them using patterns. For example:

$$3^2 = 9 \quad 3^1 = 3 \quad 3^0 = 1 \quad 3^{-1} = \frac{1}{3} \quad 3^{-2} = \frac{1}{9}$$

In this example, each time the exponent decreases by 1, the answer becomes $\frac{1}{3}$ as large. Students may also notice symmetric patterns between numbers with opposite exponents.

» *Can you use repeating decimals other than $.\overline{3}$ and $.\overline{6}$ (for example, $0.\overline{1}$)?* Yes. These can be very helpful! If you do not know what fractions they represent, you will have to explore this for a while first.

» *Can you use factorials?* Yes, you may use factorials if you like. If you do, it might lower the difficulty level of some of the cells. (Every cell can be solved without using a factorial.)

» *Can you use complex fractions?* Yes. This is a very effective strategy for solving some of the more difficult cells and for finding patterns in many of the columns.

» *Can you use negative exponents?* Yes, if you know what they mean.

Solution for Stage 2

See Solution for Stage 1.

STAGE 3

Problems for Stage 3

See grid under Stage 1. Cells labeled with a "3" are quite challenging for most people. Some of them are rarely found.

Questions and Conversations for Stage 3

» *Do you need the percent symbol? Do you ever need to use a number as a digit in a larger numeral?* No. In fact, most students never use these options. All of the cells can be solved without them, but they are available in case you want them.

» *Which cells are the very hardest?* In my experience, the two most challenging cells are $\frac{1}{4}$ with 7s, and $\frac{1}{6}$ with 8s. However, this will depend on your mathematical knowledge and hints you have been given (if any!).

» *Does every column have a pattern that works in each of its cells?* Yes, although some of the patterns are very difficult to find!

» *Are there certain strategies that are especially helpful for the most challenging cells?* Yes, using repeating decimals, square roots, and complex fractions can be very helpful, especially in combination.

» *How can you write the number* $\frac{1}{3}$ *using just a single "1"?* First, think about what $\frac{1}{9}$ looks like as a decimal. What can you do to change this to $\frac{1}{3}$? (You can take the square root of it.)

» *What is an easy way to simplify a complex fraction such as* $\dfrac{\frac{2}{7}}{\frac{3}{7}}$ *that has the same denominator in each of the two "small" fractions?* Multiply the numerator and denominator of the main fraction by the same number. (Which one?) Or rewrite the main fraction as a division expression. You will see that the answer ($\frac{2}{3}$ in the example above) comes directly from the numerators of the two "small" fractions.

Solution for Stage 3

See Solution for Stage 1.

Here are a few additional solutions that do not appear in the grid, including some that use factorials.

$\frac{1}{4}$ with ones: $\dfrac{1}{1+\left(\sqrt{\overline{.1}}\right)^{-1}}$ or $\dfrac{\sqrt{.1}\cdot 1}{\sqrt{.1}+1}$ $\frac{1}{4}$ with sixes: $\dfrac{.6}{6-(.6\cdot 6)}$ or $\dfrac{6\div .\overline{6}}{6\cdot 6}$

$\frac{1}{5}$ with twos: $\dfrac{\sqrt{2\cdot 2}}{2\div .2}$

$\frac{1}{6}$ with ones: $\dfrac{1}{(1+1+1)!}$ $\frac{1}{6}$ with eights: $\dfrac{\sqrt{8+8}}{\left(\sqrt{8+8}\right)!}$

$\frac{1}{7}$ with ones: $\left(\left(\sqrt{\overline{.1}}\right)^{-1}!+1\right)^{-1}$ $\frac{1}{7}$ with threes: $\dfrac{3}{3^3-3!}$

$\frac{1}{8}$ with twos: or $\dfrac{2}{2^{\left(2^2\right)}}$ $\frac{1}{8}$ with fives: $\dfrac{5+5+5}{5!}$

ADDITIONAL CHALLENGE

» Find a pattern for each column. Use algebra to show why the patterns work.

Solution for Additional Challenge

Patterns for the columns.

Note. For decimals, the letters represent digits rather than numbers.

The $\dfrac{1}{2}$ column: $\dfrac{\sqrt{a \cdot a}}{a+a}$ or $\dfrac{\sqrt{a} \cdot \sqrt{a}}{a+a}$

The $\dfrac{1}{3}$ column: $\dfrac{a}{a+a+a}$ (This is usually one of the first things that students notice.)

The $\dfrac{1}{4}$ column: $\dfrac{.\overline{a}+.\overline{a}}{a-.\overline{a}}$ or $\dfrac{\sqrt{.\overline{a}}}{\sqrt{.\overline{a}}+\sqrt{\sqrt{a \cdot a}}}$ (... *not* the first thing they usually notice!)

The $\dfrac{1}{5}$ column: $\dfrac{.\overline{a}+.\overline{a}}{a+.\overline{a}}$ or $\dfrac{.a}{a}+\dfrac{.a}{a}$ or $\dfrac{.a+.a}{\sqrt{a \cdot a}}$

The $\dfrac{1}{6}$ column: $\dfrac{\sqrt{a \cdot .\overline{a}}}{a+a}$ or $\dfrac{\sqrt{a} \cdot \sqrt{.\overline{a}}}{a+a}$

The $\dfrac{1}{7}$ column: $\dfrac{.\overline{a}}{a-\left(.\overline{a}+.\overline{a}\right)}$ or $\dfrac{.\overline{a}}{a-.\overline{a}-.\overline{a}}$

The $\dfrac{1}{8}$ column: $\dfrac{.a}{a-\left(.a+.a\right)}$ or $\dfrac{.a}{a-.a-.a}$

The $\dfrac{1}{9}$ column: $\dfrac{.\overline{a} \cdot a}{a \cdot a}$ or $.\overline{aaaa}$

Algebraic justifications for selected column patterns.

A pattern for $\dfrac{1}{3}$: $\dfrac{a}{a+a+a}$ $\qquad \dfrac{a}{3 \cdot a}=\dfrac{1}{3}$

A pattern for $\dfrac{1}{4}$: $\dfrac{.\overline{a}+.\overline{a}}{a-.\overline{a}}$ $\qquad \dfrac{\dfrac{a}{9}+\dfrac{a}{9}}{a-\dfrac{a}{9}}=\dfrac{\dfrac{2 \cdot a}{9}}{\dfrac{9 \cdot a}{9}-\dfrac{a}{9}}=\dfrac{\dfrac{2 \cdot a}{9}}{\dfrac{8 \cdot a}{9}}=\dfrac{\dfrac{2 \cdot a}{9} \cdot 9}{\dfrac{8 \cdot a}{9} \cdot 9}=\dfrac{2 \cdot a}{8 \cdot a}=\dfrac{2}{8}=\dfrac{1}{4}$

A pattern for $\dfrac{1}{6}$: $\dfrac{\sqrt{a \cdot .\overline{a}}}{a+a}$ $\qquad \dfrac{\sqrt{a \cdot \dfrac{a}{9}}}{a+a}=\dfrac{\sqrt{\dfrac{a}{1} \cdot \dfrac{a}{9}}}{a+a}=\dfrac{\sqrt{\dfrac{a^2}{9}}}{2 \cdot a}=\dfrac{\dfrac{a}{3}}{2 \cdot a}=\dfrac{a}{3} \cdot \dfrac{1}{2 \cdot a}=\dfrac{a}{6 \cdot a}=\dfrac{1}{6}$

A pattern for $\dfrac{1}{7}$: $\dfrac{.\overline{a}}{a-.\overline{a}-.\overline{a}}$ $\qquad \dfrac{\dfrac{a}{9}}{a-\dfrac{a}{9}-\dfrac{a}{9}}=\dfrac{\dfrac{a}{9}}{\dfrac{9 \cdot a}{9}-\dfrac{a}{9}-\dfrac{a}{9}}=\dfrac{\dfrac{a}{9}}{\dfrac{7 \cdot a}{9}}=\dfrac{\dfrac{a}{9} \cdot 9}{\dfrac{7 \cdot a}{9} \cdot 9}=\dfrac{a}{7 \cdot a}=\dfrac{1}{7}$

130

WRAP UP

Share Strategies

Have students share some of their solutions and strategies. (There will probably be many that are not included in this book!) Ask them to watch for solutions that look similar. Can they use mathematical properties to explain why the expressions have the same value? For example, you may be able to obtain one solution from another by the using the distributive property or by multiplying the numerator and denominator by the same number.

> **Teacher's Note.** If you allow students to work on this exploration until the end of the school year, you may not have the following discussions until that time!

Summarize

Answer any remaining questions that students have. You may want to share and discuss solutions to any cells that no one solved.

Further Exploration

Ask students to think of new questions to ask or ways to extend the exploration. Here are some possibilities:

» Extend the grid to include 10ths, 11ths, 12ths, etc.
» Use fractions with numerators other than 1.
» Use three copies of each whole number instead of four.
» Reverse the entire process! Try to use the unit fractions to make the whole numbers. Do any difficulties arise?

Exploration 9

Continued Fractions

INTRODUCTION

Materials

- » Graph paper
- » Calculator

Prior Knowledge

- » Know a procedure for dividing fractions.
- » Understand the concept of a reciprocal.
- » Be familiar with repeating decimals.
- » Understand the meaning of a square root (Stage 3).

Learning Goals

- » Increase confidence and computational fluency with fractions and reciprocals.
- » Compare multiple symbolic and pictorial representations of numbers, some familiar and some new. Create and justify procedures to translate between them.
- » Describe, analyze, and extend complex patterns.
- » Develop and test processes for approximating irrational numbers (Stage 3).
- » Communicate complex mathematical ideas clearly.
- » Persist in solving challenging problems.

Launching the Exploration

Motivation and purpose. To students: You know how to write numbers as fractions, decimals, and percentages, but did you know that there are other ways? The ancient Greeks, who did not have a place value system (and thus no decimals!), used ideas related to *continued fractions* to represent numbers (although they wrote them differently than we do). This system has features in common with both fractions and decimals. It also has connections to the *greatest common factor* concept and a famous number known as the *golden ratio*. As you learn about it, you will develop a deeper knowledge of reciprocals and strengthen your computational and analytical skills.

Understanding the problem. This exploration may take a little longer to introduce than others. Begin by asking students to find a simpler way to write expressions of the

133

type $\dfrac{1}{\frac{a}{b}}$. See the Questions and Conversations section in the Teacher's Guide for ideas on leading this conversation.

Once students understand why the expression is equal to $\dfrac{b}{a}$, tell them that they will be studying methods to use reciprocals to write numbers in a new form—*continued fractions*. Although these look like "layered" fractions when you write them out, you can also represent them using a bracket notation that looks similar to a decimal. For example, you can write $\dfrac{177}{41}$ (or $4.\overline{31707}$) as [4; 3, 6, 2], which means:

$$4 + \cfrac{1}{3 + \cfrac{1}{6 + \cfrac{1}{2}}}$$

Do you see the reciprocals in this expression?

Notice how the semicolon in the bracket works like a decimal point in the way that it separates the whole number from the fractional part of the number. The digits 3, 6, and 2 after the semicolon appear in the denominators of the different "fractions within fractions." The numerators are not listed in the bracket, because brackets represent *simple* continued fractions—ones whose numerators are all equal to 1. We will use only simple continued fractions in this exploration.

Look at Problem #1 with students. Ask them to describe the connection between each continued fraction and the bracket below it. Also, ask them why the digit before the semicolon is 0. (It is because there is no whole number part in the continued fraction expression).

Teacher's Note. Dr. Ron Knott (2013), former professor of mathematics and computing sciences at the University of Surrey, discusses the idea of using rectangle diagrams to represent continued fractions on his website, "Introduction to Continued Fractions" (http://www.maths.surrey.ac.uk/hosted-sites/R.Knott/Fibonacci/cfINTRO.html). His site also includes an online continued fraction calculator, continued fraction forms for square roots, and many other interesting ideas and questions to explore!

Allow students to puzzle out the process of calculating the values of the continued fractions in Problem #1. As they begin work, consider using the Questions and Conversations for #1 to guide a discussion about where and how to start.

Before students get too far into the activity, find some time to look briefly through it. In Stage 1, they become familiar with the basic processes and ideas behind continued fractions, including a way to represent them pictorially. In Stage 2, they explore and analyze patterns in continued fractions. Finally, in Stage 3, they use continued fractions to investigate irrational numbers. They will discover that, just like decimals, continued fractions can go on forever!

STUDENT HANDOUT

Stage 1

1. Write the exact value of each *simple continued fraction* in a more familiar form. Show your thinking process.

   a. $\cfrac{1}{2+\cfrac{1}{1+\cfrac{1}{3}}}$

   [0; 2, 1, 3]

   b. $\cfrac{1}{2+\cfrac{1}{3+\cfrac{1}{2}}}$

   [0; 2, 3, 2]

   c. $\cfrac{1}{1+\cfrac{1}{2+\cfrac{1}{4}}}$

   [0; 1, 2, 4]

2. Compare this diagram to Problem #1, Part A. Then create similar diagrams for 1B and 1C. Explain your thinking process, describing as many details as you can about the connections between the pictures and the continued fractions.

   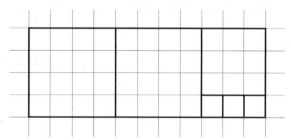

3. Find the reciprocals of the numbers in Problem #1. Show or describe what their continued fractions (in regular and bracket form) and diagrams look like. What general observations can you make regarding reciprocals and continued fractions?

4. Every continued fraction may be written in two ways—with or without a "1" at the end. Examples: [2; 5, 7] = [2; 5, 6, 1] and [0; 3, 2, 4, 1] = [0; 3, 2, 5]. Why are these continued fractions equal? What is the general rule?

5. Create at least one method (with or without using diagrams) to write fractions in continued fraction form. Use an example to illustrate your method(s). Suggested examples to try: $\frac{2}{5}$, $\frac{8}{11}$, and $\frac{24}{13}$.

6. Explore continued fraction forms of fractions with denominators of 5: $\frac{1}{5}$, $\frac{2}{5}$, $\frac{3}{5}$, and $\frac{4}{5}$. Then look at fractions with numerators of 5: $\frac{5}{6}$, $\frac{5}{7}$, $\frac{5}{8}$, $\frac{5}{9}$, $\frac{5}{10}$, $\frac{5}{11}$, etc. Describe any patterns you see. Explain what causes some of the patterns.

7. Find formulas for writing the continued fractions $[0; a, b]$ and $[0; a, b, c]$ as regular fractions. Can you find patterns between the formulas?

Continued fractions can go on forever, just like decimals! For example, the notation $[1; \overline{2}]$ means that the digit 2 repeats forever.

$$[1; 2, 2, 2, 2, 2, 2, \ldots]$$

In Stage 3, you explore these kinds of continued fractions. You will need to know a few facts about *rational* and *irrational* numbers. Saying that a decimal or continued fraction *terminates* means that it eventually ends.

	Decimal	Continued Fraction
Rational Number	Always terminates or repeats	Always terminates
Irrational Number	Never terminates or repeats	Never terminates, but may repeat

Now the square root of a whole number is either a whole number or an irrational number. In the latter case, the decimal never terminates or repeats, but its continued fraction form does repeat!

8. Explore the infinite continued fraction $[1; \overline{2}]$. Suggestion: Calculate the values of $[1; 2]$, $[1; 2, 2]$, $[1; 2, 2, 2]$, $[1; 2, 2, 2, 2]$, etc. (These are called the *convergents* of the continued fraction.) Look for patterns in your results and try to extend them. Each time you put another 2 in the bracket, the result gets closer to a certain square root. Which one? How can you tell?

9. Carry out the same type of investigation for the continued fraction $[2; \overline{4}]$.

TEACHER'S GUIDE

INTRODUCTION

Questions and Conversations for the Introduction

This section contains ideas for conversations, mainly in the form of questions that students may ask or that you may pose to them. Be sure to allow students to do most of the thinking and talking!

Below are several questions for developing ideas to write $\dfrac{1}{\frac{3}{7}}$ in a simpler form.

» *Does it help to remember the connection between fractions and division?* Yes. Think of the expression as $1 \div \dfrac{3}{7}$. From here, there are multiple possibilities. For example, you might use the standard procedure for dividing fractions (i.e., rewrite it as $1 \cdot \dfrac{7}{3}$). Or you could ask how many groups of $\dfrac{3}{7}$ are in 1 whole.

» *What is a familiar procedure for simplifying fractions? Will it apply here?* A common method is to multiply or divide the numerator and denominator by the same nonzero number. Which whole number would be the most helpful? (Answer: 7)

» *What is the definition of a reciprocal? How can you apply it here?* Reciprocals are pairs of numbers whose product is 1. This implies that 1 divided by any number (except 0) is its reciprocal. Or, because $\dfrac{3}{7} \cdot \dfrac{7}{3} = 1$, you can make the denominator of $\dfrac{1}{\frac{3}{7}}$ equal to 1 by multiplying the numerator and denominator of the main fraction by $\dfrac{7}{3}$.

» *Can you generalize your results to other expressions of the form $\dfrac{1}{\frac{a}{b}}$?* Yes, the expression can be written as $\dfrac{b}{a}$ (the reciprocal of $\dfrac{a}{b}$).

STAGE 1

Problem #1

1. Write the exact value of each *simple continued fraction* in a more familiar form. Show your thinking process.

a.
$$\cfrac{1}{2+\cfrac{1}{1+\cfrac{1}{3}}}$$

b.
$$\cfrac{1}{2+\cfrac{1}{3+\cfrac{1}{2}}}$$

c.
$$\cfrac{1}{1+\cfrac{1}{2+\cfrac{1}{4}}}$$

$[0; 2, 1, 3]$ $[0; 2, 3, 2]$ $[0; 1, 2, 4]$

Questions and Conversations for #1

» *Where do you begin the calculation? Why?* You need to begin at the bottom and work your way up. If you tried to begin at the top, you would have to know the value of the entire denominator of the main fraction, in which case you would be forced to move toward the bottom of the expression anyway.

» *Does the length of each fraction bar matter in a continued fraction?* Yes, it does. It helps determine the grouping of the expressions inside the continued fraction. For example, $\dfrac{1}{\dfrac{3}{5}}$ means $1 \div \dfrac{3}{5}$, while $\dfrac{\dfrac{1}{3}}{5}$ means $\dfrac{1}{3} \div 5$. These expressions are not equal.

» *How can you write the expression in Question 1(A) using parentheses?* It looks like this:

$$1 \div \left(2 + 1 \div \left(1 + \left(1 \div 3\right)\right)\right).$$

This will guide you through the calculation. Working with the inner set of parentheses first is equivalent to starting at the bottom of the continued fraction.

> **Teacher's Note.** Some students write their answers as decimals, especially if they use calculators. If so, they should also do the calculations with fractions and without a calculator. Take this opportunity to (1) compare the fraction and decimal forms of the answers, (2) point out the "reciprocal" key on a scientific calculator (usually $\dfrac{1}{x}$ or x^{-1}), and (3) talk about how fraction bars work as grouping symbols.

Solution for #1

The answers are a. $\dfrac{4}{11}$ or $0.\overline{36}$, b. $\dfrac{7}{16}$ or 0.4375, and c. $\dfrac{9}{13}$ or $0.\overline{692307}$.

You can calculate each of these beginning at the bottom and working up.

a.
$$\cfrac{1}{2+\cfrac{1}{1+\cfrac{1}{3}}} = \cfrac{1}{2+\cfrac{1}{\frac{4}{3}}} = \cfrac{1}{2+\frac{3}{4}} = \cfrac{1}{\frac{11}{4}} = \frac{4}{11}$$

b. $\dfrac{1}{2+\dfrac{1}{3+\dfrac{1}{2}}} = \dfrac{1}{2+\dfrac{1}{\frac{7}{2}}} = \dfrac{1}{2+\dfrac{2}{7}} = \dfrac{1}{\frac{16}{7}} = \dfrac{7}{16}$

c. $\dfrac{1}{1+\dfrac{1}{2+\dfrac{1}{4}}} = \dfrac{1}{1+\dfrac{1}{\frac{9}{4}}} = \dfrac{1}{1+\dfrac{4}{9}} = \dfrac{1}{\frac{13}{9}} = \dfrac{9}{13}$

Problem #2

2. Compare this diagram to Problem #1, Part A. Then create similar diagrams for 1B and 1C. Explain your thinking process, describing as many details as you can about the connections between the pictures and the continued fractions.

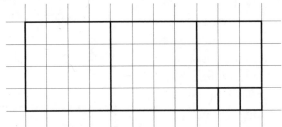

Questions and Conversations for #2

» *What key features could you use to describe the diagram?* Important features of the diagram are the length and width of the rectangle and the fact that it is subdivided into squares. How many squares of each size are there?

» *What are the key features that define the continued fraction?* Because the numerators are all equal to 1, focus on the numbers along the left side of the continued fraction (the numbers in the bracket).

Solution for #2

The numbers in the bracket tell you how many squares of each size are in the rectangle (beginning with the largest). The numerator and denominator of the answer correspond to the dimensions of the rectangle.

b. 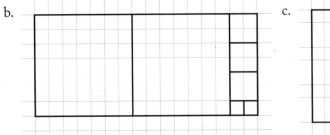   c.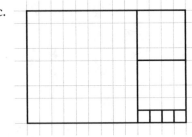

Sample strategy 1: Use the numbers in the bracket to draw connected squares so that they form a rectangle. Begin with 1 by 1 squares. For example, in Part C, begin with the number 4 at the right end of the bracket [0; 1, 2, 4]. Draw four 1 by 1 squares so that they make a 1 by 4 rectangle. Now, using the number 2 in the bracket, join two larger (4 by 4) squares to the longer side of the rectangle. Finally, moving to the number 1 in the bracket, attach one 9 by 9 square to the longer side of the resulting rectangle. This creates a 9 by 13 rectangle, corresponding to the answer $\frac{9}{13}$.

Sample strategy 2: Draw the rectangle using the answer and then subdivide it using the largest possible square at each step. For example, because the answer to Part C is $\frac{9}{13}$, draw a 9 by 13 rectangle. The largest square you can draw inside it is 9 by 9. This leaves a 4 by 9 rectangle, which holds two 4 by 4 squares. Finally, split the remaining 1 by 4 rectangle into four 1 by 1 squares. The 1 large, 2 medium, and 4 small squares correspond to the numbers in the bracket: [0; 1, 2, 4].

You can use Strategy 1 to write a continued fraction as a single simpler fraction, while you can use Strategy 2 in reverse—to write a fraction as a continued fraction.

Problem #3

3. Find the reciprocals of the numbers in Problem #1. Show or describe what their continued fractions (in regular and bracket form) and diagrams look like. What general observations can you make regarding reciprocals and continued fractions?

Questions and Conversations for #3

» *Why won't the continued fractions of the reciprocals have a "0" before the semi-colon?* The original fractions are less than 1, so their reciprocals are greater than 1. The number before the semicolon is the whole number part of the reciprocal.

» *What is the reciprocal of* $\frac{1}{x}$ *? How does this relate to the continued fraction?* The reciprocal of $\frac{1}{x}$ is x. Think of x as the entire expression in the denominator of the main fraction.

Solution for #3

The reciprocals of the answers to Problem #1 are $\frac{11}{4}$, $\frac{16}{7}$, and $\frac{13}{9}$.

You can find their continued fractions by taking the denominators of the original continued fractions.

a.
$$2+\cfrac{1}{1+\cfrac{1}{3}}$$

b.
$$2+\cfrac{1}{3+\cfrac{1}{2}}$$

c.
$$1+\cfrac{1}{2+\cfrac{1}{4}}$$

Teacher's Note. To understand why

$2+\cfrac{1}{1+\cfrac{1}{3}}$ is the reciprocal of $\cfrac{1}{2+\cfrac{1}{1+\cfrac{1}{3}}}$,

think of $2+\cfrac{1}{1+\cfrac{1}{3}}$ as x in the expression $\dfrac{1}{x}$.

In bracket notation, these are [2; 1, 3], [2; 3, 2], and [1; 2, 4]. You obtain them by shifting the digits in the original bracket one place to the left! (And you can find the reciprocals of these by shifting the digits back to where they were.)

The pictures for the reciprocals are identical to the pictures for the original numbers! You simply interpret them differently. Numbers between 0 and 1 have "0" to the left of the semicolon, while numbers greater than 1 have their whole number part in this position.

Problem #4

4. Every continued fraction may be written in two ways—with or without a "1" at the end. Examples: [2; 5, 7] = [2; 5, 6, 1] and [0; 3, 2, 4, 1] = [0; 3, 2, 5]. Why are these continued fractions equal? What is the general rule?

Teacher's Note. If students are not sure how to begin, suggest that they write the continued fractions out in detail and begin calculating their values.

Solution for #4

You can see that the expressions are equal by writing out the continued fractions. For example, [2; 5, 7] = [2; 5, 6, 1] because

$$2+\cfrac{1}{5+\cfrac{1}{7}}=2+\cfrac{1}{5+\cfrac{1}{6+\cfrac{1}{1}}}\cdot$$

This works since $7=6+\dfrac{1}{1}$.

The general rule is that if a continued fraction does not have a "1" at the end, you can subtract 1 from the last digit and insert "1" at the end. If it does have a "1" at the end, remove it and add it to the final digit.

STAGE 2

Problem #5

5. Create at least one method (with or without using diagrams) to write fractions in continued fraction form. Use an example to illustrate your method(s). Suggested examples to try: $\frac{2}{5}$, $\frac{8}{11}$, and $\frac{24}{13}$.

Questions and Conversations for #5

» *How can you think of a 2 by 5 rectangle as representing either* $\frac{2}{5}$ *or* $\frac{5}{2}$ *of a square?* It is $\frac{2}{5}$ of a 5 by 5 square, and it is $\frac{5}{2}$ of a 2 by 2 square.

» *How does the process of subdividing a rectangle with squares relate to division?* When you think of the rectangle as representing an improper fraction, subdividing it in this fashion shows the rectangle as a mixed number. The squares represent its whole number part, and the remaining smaller rectangle represents the fractional part.

» *How can you write a fraction like* $\frac{2}{5}$ *as an equivalent expression using its reciprocal?* Think about using a complex fraction. (For example, $\frac{1}{\frac{5}{2}}$)

> **Teacher's Note.** If students are having trouble devising a procedure, refer them to Problem #2. Consider drawing the rectangle and decomposing it into squares. (See Sample Strategy 2 in the Solutions.) Alternatively, study the procedure in Problem #1. Find a way to reverse the process.

Solution for #5

The answers for the three suggested numbers are:

$$\frac{2}{5} = [0; 2, 2] \qquad \frac{8}{11} = [0; 1, 2, 1, 2] \qquad \frac{24}{13} = [1; 1, 5, 2]$$

We will illustrate two methods with the fraction $\frac{8}{11}$.

Strategy 1: Use diagrams. Draw an 8 by 11 rectangle and subdivide it using the largest squares possible at each step. We will show the process in stages.

 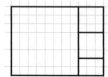 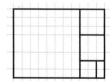 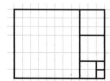

The numbers of each type of square from largest to smallest are 1, 2, 1, and 2, so the bracket notation for the continued fraction is [0; 1, 2, 1, 2]. Begin the bracket notation with 0, because the fraction is less than 1.

Strategy 2: Use numbers. Whenever a proper fraction appears, rewrite it as "1 over its reciprocal." When working with an improper fraction, rewrite it as a mixed number (using the "+" symbol). Continue until all numerators are 1.

$$\frac{8}{11} = \frac{1}{\frac{11}{8}} = \frac{1}{1+\frac{3}{8}} = \frac{1}{1+\frac{1}{\frac{8}{3}}} = \frac{1}{1+\frac{1}{2+\frac{2}{3}}} = \frac{1}{1+\frac{1}{2+\frac{1}{\frac{3}{2}}}} = \frac{1}{1+\frac{1}{2+\frac{1}{1+\frac{1}{2}}}}$$

Teacher's Note. When students share strategies, discuss the connections between the two methods. For example, when they draw the 8 by 8 square, they are thinking of the original rectangle as $\frac{11}{8}$ of it and then decomposing it into 1 whole plus $\frac{3}{8}$ of the whole. When they write $\frac{3}{8}$ as $\frac{1}{\frac{8}{3}}$ in the next step, they are reimagining the rectangle as $\frac{8}{3}$ of a 3 by 3 square.

Problem #6

6. Explore continued fraction forms of fractions with denominators of 5: $\frac{1}{5}$, $\frac{2}{5}$, $\frac{3}{5}$, and $\frac{4}{5}$. Then look at fractions with numerators of 5: $\frac{5}{6}$, $\frac{5}{7}$, $\frac{5}{8}$, $\frac{5}{9}$, $\frac{5}{10}$, $\frac{5}{11}$, etc. Describe any patterns you see. Explain what causes some of the patterns.

Questions and Conversations for #6

» *How can you organize your work to make patterns more visible?* Write the fractions in order with their continued fraction representations right next to them. If you begin to see patterns, adjust your organization to highlight them. Using a table may help.

» *What connection is suggested by the fact that some fractions have 5 in the numerator while the others have 5 in the denominator?* There may be a connection to reciprocals.

Solution for #6

Denominators of 5: $1 = [0;5]$, $\frac{2}{5} = [0;2,2]$, $\frac{3}{5} = [0;1,1,2]$, $\frac{4}{5} = [0;1,4]$

Numerators of 5:

$$\frac{5}{5}=[0;1] \qquad \frac{5}{10}=[0;2] \qquad \frac{5}{15}=[0;3]$$

$$\frac{5}{6}=[0;1,5] \qquad \frac{5}{11}=[0;2,5] \qquad \frac{5}{16}=[0;3,5]$$

$$\frac{5}{7}=[0;1,2,2] \qquad \frac{5}{12}=[0;2,2,2] \qquad \frac{5}{17}=[0;3,2,2]$$

$$\frac{5}{8}=[0;1,1,1,2] \qquad \frac{5}{13}=[0;2,1,1,2] \qquad \frac{5}{18}=[0;3,1,1,2]$$

$$\frac{5}{9}=[0;1,1,4] \qquad \frac{5}{14}=[0;2,1,4] \qquad \frac{5}{19}=[0;3,1,4]$$

$\frac{5}{5}$ is shown as [0; 1] instead of [1; 0] to better fit the patterns in the table.

Fractions with numerators of 5 have continued fraction representations that fall into groups of 5 based on the first digit after the semicolon. The rest of the digits correspond to the digits in one of the continued fractions for $\frac{1}{5}$, $\frac{2}{5}$, $\frac{3}{5}$, and $\frac{4}{5}$.

Test this by taking the reciprocals of the fractions with the numerators of 5.

For example, $\frac{5}{7}$ has a reciprocal of $\frac{7}{5}$ or $1\frac{2}{5}$. The continued fraction for $1\frac{2}{5}$, ([1; 2, 2]), looks like the one for $\frac{2}{5}$ ([0; 2, 2]) except that it has a "1" before the semicolon. To return to $\frac{5}{7}$, shift the digits in the bracket one place to the right, giving [0; 1, 2, 2]. This puts the "1" right after the semicolon, followed by the digits for $\frac{2}{5}$.

Some students may observe that increasing the digit after the semicolon by 1 adds 5 (the numerator) to the denominator while leaving the numerator unchanged.

Problem #7

7. Find formulas for writing the continued fractions [0; a, b] and [0; a, b, c] as regular fractions. Can you find patterns between the formulas?

Questions and Conversations for #7

» *What types of strategies are available to you?* Possibilities include:
- Looking for patterns in numerical examples.
- Drawing the rectangle diagrams.
- Using algebra.

Solution for #7

The formulas can be written $[0; a, b] = \dfrac{b}{a \cdot b + 1}$ and $[0; a, b, c] = \dfrac{b \cdot c + 1}{a \cdot b \cdot c + a + c}$.

Most students will probably search for these formulas by creating numerical examples and looking for patterns. Some may be able to carry out algebraic procedures. For example:

$$[0; a, b] = \cfrac{1}{a + \cfrac{1}{b}} = \cfrac{1}{a + \cfrac{1}{b}} \cdot \frac{b}{b} = \frac{b}{a \cdot b + 1}$$

(The algebraic approach is much more complicated for $[0; a, b, c]$!)

A few students may use diagrams. For example, to find a formula for $[0; a, b]$, begin by joining b 1 by 1 squares to make a 1 by b rectangle.

Then join a squares to the longer side of the rectangle.

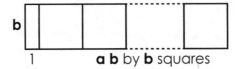

This rectangle has a length and width of b and $a \cdot b + 1$, resulting in the fraction $\dfrac{b}{a \cdot b + 1}$.

To find the formula for $[0; a, b, c]$, change the labels from a to b and b to c. Then imagine joining a squares that have side lengths of $b \cdot c + 1$ to the picture above.

STAGE 3

Continued fractions can go on forever, just like decimals! For example, the notation $[1; \overline{2}]$ means that the digit 2 repeats forever.

$[1; 2, 2, 2, 2, 2, 2, \ldots]$

In Stage 3, you explore these kinds of continued fractions. You will need to know a few facts about *rational* and *irrational* numbers. Saying that a decimal or continued fraction *terminates* means that it eventually ends.

	Decimal	Continued Fraction
Rational Number	Always terminates or repeats	Always terminates
Irrational Number	Never terminates or repeats	Never terminates, but may repeat

Now, the square root of a whole number is either a whole number or an irrational number. In the latter case, the decimal never terminates or repeats, but its continued fraction form does repeat!

Problem #8

8. Explore the infinite continued fraction $[1; \overline{2}]$. Suggestion: Calculate the values of $[1; 2]$, $[1; 2, 2]$, $[1; 2, 2, 2]$, $[1; 2, 2, 2, 2]$, etc. (These are called the *convergents* of the continued fraction.) Look for patterns in your results and try to extend them. Each time you put another 2 in the bracket, the result gets closer to a certain square root. Which one? How can you tell?

Teacher's Note. If students are not familiar with the definitions of rational and irrational numbers, begin by defining a *rational number* as a number that can be written as a terminating or repeating decimal. (This is usually the easiest form of the definition for students to understand at first.) Irrational numbers are defined as those that cannot be written in either of these forms. As students work through Stage 3, they will begin to see that irrational numbers cannot be written exactly as simple fractions (fractions with whole number numerators and denominators) either!

Solution for #8

The values of the first few convergents and their decimal forms are:

$$[1; 2] = 1\frac{1}{2} \text{ or } \frac{3}{2} \qquad 1.5$$

$$[1; 2, 2] = 1\frac{2}{5} \text{ or } \frac{7}{5} \qquad 1.4$$

$$[1; 2, 2, 2] = 1\frac{5}{12} \text{ or } \frac{17}{12} \qquad 1.41\overline{6}$$

$$[1; 2, 2, 2, 2] = 1\frac{12}{29} \text{ or } \frac{41}{29} \qquad 1.413793103448\ldots$$

The convergents are getting closer to $\sqrt{2}$. One way to see this is to square them. We will show both the fraction and decimal forms.

$$\left(\frac{3}{2}\right)^2 = \frac{9}{4} \qquad 1.5^2 = 2.25$$

$$\left(\frac{7}{5}\right)^2 = \frac{49}{25} \qquad 1.4^2 = 1.96$$

$$\left(\frac{17}{12}\right)^2 = \frac{289}{144} \qquad 1.41\overline{6}^2 = 2.0069\overline{4}$$

$$\left(\frac{41}{29}\right)^2 = \frac{1681}{841} \qquad 1.413793103448^2 \approx 1.99881093936$$

The squares of the convergents are getting closer and closer to 2 (quickly!). This is very clear in the fractions $\frac{9}{4}$, $\frac{49}{25}$, $\frac{289}{144}$, and $\frac{1681}{841}$. The numerators are alternately 1 greater and 1 less than twice the denominator!

$$9 = 4 \cdot 2 + 1 \qquad 49 = 25 \cdot 2 - 1 \qquad 289 = 144 \cdot 2 + 1 \qquad 1681 = 841 \cdot 2 - 1$$

Students may also notice patterns in the convergents. For the fractional part of the mixed numbers ($\frac{1}{2}$, $\frac{2}{5}$, $\frac{5}{12}$, $\frac{12}{29}$), you can predict the next convergent using the pattern:

$$\frac{a}{b} \rightarrow \frac{b}{a + 2 \cdot b}.$$

For the improper fractions ($\frac{3}{2}$, $\frac{7}{5}$, $\frac{17}{12}$, $\frac{41}{29}$), the pattern looks like this:

$$\frac{a}{b} \rightarrow \frac{a + 2 \cdot b}{a + b}.$$

Based on this, students can predict and test more convergents: $\frac{99}{70}$, $\frac{239}{169}$, $\frac{577}{408}$.

Problem #9
- - - - - - - - - - - -
9. Carry out the same type of investigation for the continued fraction $[2; \overline{4}]$.

Solution for #9

This infinite continued fraction equals $\sqrt{5}$ or approximately 2.2360679775. Below is a summary of data students may gather when they investigate.

Continued Fraction	Mixed Number and Fraction, $\frac{x}{y}$	Decimal	$\left(\frac{x}{y}\right)^2$	Relationship Between Numerator and Denominator
[2; 4]	$2\frac{1}{4}$ or $\frac{9}{4}$	2.25	$\frac{81}{16}$	$81 = 16 \cdot 5 + 1$
[2; 4, 4]	$2\frac{4}{17}$ or $\frac{38}{17}$	2.23529411765 ...	$\frac{1444}{289}$	$1444 = 289 \cdot 5 - 1$
[2; 4, 4, 4]	$2\frac{17}{72}$ or $\frac{161}{72}$	$2.236\overline{1}$	$\frac{25{,}921}{5184}$	$25{,}921 = 5184 \cdot 5 + 1$
[2; 4, 4, 4, 4]	$2\frac{72}{305}$ or $\frac{682}{305}$	2.23606557377 ...	$\frac{465{,}124}{93{,}025}$	$465{,}124 = 93{,}025 \cdot 5 - 1$

Sample observations:

» The fraction part of the mixed numbers ($\frac{1}{4}$, $\frac{4}{17}$, $\frac{17}{72}$, $\frac{72}{305}$, etc.) follows the pattern:

$$\frac{a}{b} \rightarrow \frac{b}{a + 4 \cdot b}.$$

» The convergents get closer and closer to $\sqrt{5}$ very quickly, alternately going above and below it, just as before.

» The numerator of the square of the convergents is alternately 1 greater and 1 less than 5 times its denominator, illustrating just how close the value of the fraction is to 5!

» You can use these patterns to predict the values of more convergents:

$$2\frac{305}{1292}, \ 2\frac{1292}{5473}, \text{ etc.}$$

149

WRAP UP

Share Strategies

Have students compare and critique one another's strategies and observations. Use this opportunity to identify and correct any misconceptions.

Summarize

Answer any remaining questions that students have. You may also want to summarize and expand on a few key ideas:

» Simple continued fractions represent a process of alternately adding and taking reciprocals.

» Fraction bars often serve as grouping symbols. It is important to write them carefully to make the intended grouping clear.

» Continued fractions create many interesting patterns to explore.

» You can write rational numbers in exact form as simple fractions, or as terminating or repeating decimals. You cannot write irrational numbers exactly in either decimal or simple fraction form.

» Even though irrational numbers may have decimals that appear unpredictable, you can sometimes use predictable processes to approximate them.

» The continued fraction form of a number is often simpler than its decimal form. For example, repeating decimals can always be written as terminating continued fractions, and some irrational numbers can be written as repeating continued fractions.

» An interesting fact: A convergent of a continued fraction gives you the best possible rational approximation of an irrational number for any denominator of equal or lesser value. For example, $\frac{49}{25}$ is the closest possible rational approximation of $\sqrt{2}$ using a denominator less than or equal to 25.

Further Exploration

Ask students to think of new questions to ask or ways to extend this exploration. Here are some possibilities:

» Research and explore connections between continued fractions, Fibonacci numbers, and the golden ratio.

» Continue exploring patterns similar to those in Problem #6. Choose numerators and denominators other than 5.

» What happens when you reverse the order of the numbers in a bracket—for example, when you change [0; 1, 3, 2] to [0; 2, 3, 1]? (Remember that there are two ways to write every continued fraction.)

» Find the continued fraction forms of more square roots. You will see many complex and beautiful patterns!

» Did you notice that the fractions in this activity were all in simplest form? What is causing this? (Do some research on connections to the Euclidean Algorithm.)

» Continue searching for formulas for continued fractions. Look for patterns! (Sample answer: $[0; a, b, c, d] = \dfrac{bcd + b + d}{abcd + ab + ad + cd + 1}$)

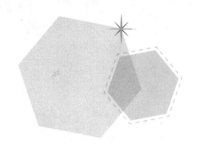

References

Knott, R. (2013, October 1). *An introduction to continued fractions*. Retrieved from http://www.maths.surrey.ac.uk/hosted-sites/R.Knott/Fibonacci/cfINTRO.html

National Governors Association Center for Best Practices, & Council of Chief State School Officers. (2010). Common core state standards for mathematics. Washington, DC: Authors.

Sheffield, L. J. (2003). *Extending the challenge in mathematics: Developing mathematical promise in K–8 students*. Thousand Oaks, CA: Corwin Press.

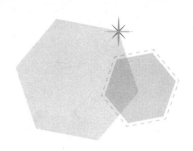

About the Author

Jerry Burkhart has been teaching and learning math with gifted students in Minnesota for nearly 20 years. He has degrees in physics, mathematics, and math education from University of Colorado, Boulder, and Minnesota State University, Mankato. Jerry provides professional development for teachers and is a regular presenter at conferences addressing topics of meeting the needs of gifted students in mathematics.

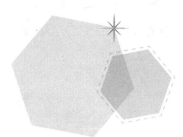

Common Core State Standards Alignment

Exploration	Common Core State Standards in Mathematics
Exploration 1: Sharing and Grouping	5.NF.B Apply and extend previous understandings of multiplication and division. 7.RP.A Analyze proportional relationships and use them to solve real-world and mathematical problems. 6.NS.A Apply and extend previous understandings of multiplication and division to divide fractions by fractions.
Exploration 2: Fraction Puzzlers	5.NF.A Use equivalent fractions as a strategy to add and subtract fractions. 6.NS.C Apply and extend previous understandings of numbers to the system of rational numbers. 7.NS.A Apply and extend previous understandings of operations with fractions. 4.NF.A Extend understanding of fraction equivalence and ordering.
Exploration 3: Working Together	5.NF.A Use equivalent fractions as a strategy to add and subtract fractions. 6.EE.A Apply and extend previous understandings of arithmetic to algebraic expressions. 7.EE.B Solve real-life and mathematical problems using numerical and algebraic expressions and equations. 6.RP.A Understand ratio concepts and use ratio reasoning to solve problems. 5.MD.A Convert like measurement units within a given measurement system.
Exploration 4: Fractions Forever!	5.NF.A Use equivalent fractions as a strategy to add and subtract fractions. 7.EE.B Solve real-life and mathematical problems using numerical and algebraic expressions and equations. 5.OA.A Write and interpret numerical expressions. 5.OA.B Analyze patterns and relationships. 7.NS.A Apply and extend previous understandings of operations with fractions. 8.NS.A Know that there are numbers that are not rational, and approximate them by rational numbers.

Exploration	Common Core State Standards in Mathematics
Exploration 5: Visualizing Fraction Multiplication	5.NF.B Apply and extend previous understandings of multiplication and division.
Exploration 6: Undo It!	6.NS.A Apply and extend previous understandings of multiplication and division to divide fractions by fractions. 5.NF.B Apply and extend previous understandings of multiplication and division. 6.RP.A Understand ratio concepts and use ratio reasoning to solve problems. 7.RP.A Analyze proportional relationships and use them to solve real-world and mathematical problems.
Exploration 7: Sum-Product Pairs	7.NS.A Apply and extend previous understandings of operations with fractions. 7.EE.B Solve real-life and mathematical problems using numerical and algebraic expressions and equations.
Exploration 8: Unit Fraction Hunt	7.NS.A Apply and extend previous understandings of operations with fractions. 7.EE.A Use properties of operations to generate equivalent expressions. 7.EE.B Solve real-life and mathematical problems using numerical and algebraic expressions and equations. 8.EE.A Work with radicals and integer exponents.
Exploration 9: Continued Fractions	7.NS.A Apply and extend previous understandings of operations with fractions. 8.NS.A Know that there are numbers that are not rational, and approximate them by rational numbers. 8.EE.A Work with radicals and integer exponents. 7.EE.A Use properties of operations to generate equivalent expressions. 7.EE.B Solve real-life and mathematical problems using numerical and algebraic expressions and equations.

Note: Please see p. 8 of the book for details on how to connect and extend the core learning of content in these lessons.